W9-DFE-407

D1274727

GOING TO TEXAS
FIVE CENTURIES OF TEXAS MAPS

TEXAS MAPS
FROM THE
MUSEUM OF THE BIG BEND
SUL ROSS UNIVERSITY

YANA AND MARTY DAVIS MAP COLLECTION

PRESENTED BY
THE CENTER FOR TEXAS STUDIES AT TCU

———◆———

TCU PRESS
FORT WORTH, TEXAS

Copyright © The Center for Texas Studies at TCU, 2007

Library of Congress Cataloging-in-Publication Data

Texas Christian University. Center for Texas Studies.
Going to Texas, five centuries of Texas maps / presented by the Center for Texas Studies at TCU
p. cm.
includes bibliographical references and index.
ISBN 978-0-87565-344-0
Texas—Historical geography—Maps. 1. Title.

G1371.S1.T4 2007
911'.764   2007061737

C O V E R   M A P :   P L A T E   5 1

Map of Texas, Arbuckle Bros. Coffee Company, published by John and Charles Arbuckle, Pittsburgh.
Chromolithograph by Donaldson Bros., New York, 1889. Trade card promotional, 3" x 5".

*Pittsburgh grocers John and Charles Arbuckle patented a method for roasting coffee involving coating
beans with a glaze of egg and sugar to seal in the flavor. They marketed their popular product under the
name of "Arbuckles' Ariosa coffee." The one-pound, airtight packages became a standard on cattle drives.
The coffee was freighted to ranches in wooden crates containing one hundred packages. Each crate
included an assortment of coupons redeemable for notions that varied from straight razors to cowboy
wedding rings. Also included were peppermint candy and trade cards such as this map.*

Cover and text design by Whittington & Co., Austin, Texas

Printed in China

Contents

PLATES

ACKNOWLEDGEMENTS

The Center for Texas Studies at TCU wishes to acknowledge the contributions of Ben Huseman, Cartographic Archivist at the University of Texas at Arlington, for his efforts to ensure that all map attributions in this volume are correct. It also wishes to thank the Lesley Family Foundation, the Burnett Foundation, Clifton and Shirley Caldwell, the Amon G. Carter Foundation and the Summerlee Foundation for generously contributed funds to underwrite this exhibition.

CONTRIBUTORS

Gregg Cantrell holds the Erma and Ralph Lowe Chair in Texas History at TCU. He has written widely on topics in nineteenth-century Texas history. His books include *Stephen F. Austin, Empresario of Texas* and *The History of Texas* (3rd ed.) which he coauthored with Robert A. Calvert and Arnoldo De Leon.

Donald E. Chipman is emeritus professor of history at the University of North Texas. He is the author or coauthor of five books relating to the history of Colonial Mexico and Spanish Texas. In 2003 Chipman was inaugurated into the Order of Isabella the Catholic as a lord/knight on orders of King Juan Carlos I of Spain.

David Coffey holds a Ph.D. from TCU and currently serves as interim dean of the college of humanities and fine arts and chair of the department of history and philosophy at the University of Tennessee at Martin. He is the author of several books and articles, including *Sheridan's Lieutenants: Phil Sheridan, His Generals, and the Final Year of the Civil War*, and was an editor of the award-winning *Encyclopedia of the Vietnam War* and the *Encyclopedia of American Military History*.

Marty Davis holds a J.D. degree from Southern Methodist University and is a C.P.A. heading Davis, Clark and Company in Dallas. History is his avocation.

Lawrence John Francell has a B.A. in history from Austin College and an M.A. in history from the University of Texas at Austin. He is the director of the Museum of the Big Bend, Sul Ross State University, Alpine, and author of *Fort Lancaster: Texas Frontier Sentinel*.

Robert Maberry Jr. has taught history at Texas Christian University. He is a former guest curator at the Museum of Fine Arts, Houston, and author of the award-winning book *Texas Flags*.

Rebecca Sharpless is assistant professor of history at Texas Christian University and the author of *Fertile Ground, Narrow Choices: Women on Texas Cotton Farms, 1900-1940*. From 1993 to 2006, she directed the Baylor University Institute for Oral History.

Gene Allen Smith is professor of history and director of the Center for Texas Studies at TCU. He has written, edited, or co-authored several books on aspects of naval and maritime history, the War of 1812, and early American territorial expansion in the Gulf South, including *Filibusters and Expansionists: Jeffersonian Manifest Destiny, 1800-1821* (1997), and *Thomas ap Catesby Jones: Commodore of Manifest Destiny* (2000).

Matt Walter is curator at the Museum of the Big Bend in Alpine, Texas. Retired from the U.S. Coast Guard in 1996, he moved to the Big Bend region of Texas, where he worked as a seasonal park ranger while completing his master's in history at Sul Ross State University in 2002.

Richard Bruce Winders is a recognized authority on nineteenth-century conflicts between the U.S. and Mexico. He holds the position of historian and curator at the Alamo.

GOING TO TEXAS

MARTY DAVIS

After a quarter century of collecting Texas maps, I have concluded that an unspoken assumption exists among collectors. Most believe that the reason for exploration of the New World was a search for Texas. Columbus stopped short of that discovery. Cabeza de Vaca (1529-1536), Coronado (1541), and the remnants of De Soto's expedition (1542) took years to waltz across Texas. La Salle claimed to have mistaken Matagorda Bay for the Mississippi Delta, and he, too, landed on Texas soil. All such explorers left home with no better itinerary than seeking "parts unknown." Whether they were motivated by adventure, treasure or religion, the result was the same. They left the country they visited as they found it—untamed and uncharted.

During the latter part of the nineteenth century, the initials "G.T.T." or "Gone to Texas" were the standard farewell given before striking out for Texas. Newcomers to the state sought the promised land. Whether navigated by star-based *portolanos* or satellite-driven global positioning devices, the arduous trek seemed worthwhile. The lure of the "Great Space of Land Unknown" lured thousands to the pilgrimage. "Gone to Texas" is the cartographic legacy of those sojourners.

The inspiration for this project was an Amon Carter Museum exhibit twenty years ago, "Crossroads of Empire." The standard work on Texas cartography, *Maps of Texas and the Southwest: 1513-1900*, was a consequence of that effort. The book's dust jacket said that the fifty maps included in the exhibition represented "every historically significant map of Texas," and the Museum of the Big Bend map collection contains most of the items that were in that show. In writing *Maps of Texas*, authors James and Robert Martin aimed "to stimulate further inquiry and to add to the appreciation of the value of historic maps as tools in learning about any place, in any time."

That goal was accomplished, and, since that time, more map collections have been donated to universities, such as the Virginia Garrett Cartographic Collection at the University of Texas at Arlington and the Barr-Rowe Collection at Southern Methodist University's De Golyer Library. With the help of private donors, the Texas General Land Office has also restored many of its original maps.

The "Going to Texas: Five Centuries of Texas Maps" exhibit differs from its progenitor in several aspects. The maps displayed come from a single collection rather than from several institutions, and the exhibition is a collaborative effort between the Center for Texas Studies at Texas Christian University, the Museum of the Big Bend, and eleven regional museums. Finally, the time frame represented extends one hundred years beyond the Amon Carter exhibition.

Ancient cartographers strove to combine art and science to give "graphic illustration" to geography. Knowledge, tradition, politics, and economics all contributed in some degree to the final product. This exhibit groups Texas maps into five segments. Early maps of discovery were crude and fanciful and gave only an impression of the actual geography. After Columbus, subsequent adventures explored deeper into the mainland. Gastaldi's Atlas of 1548 (Plate 1), devoted exclusively to New Spain, was based on interior exploration. Forty years later Ortelius detailed the Gulf of Mexico (Plate 2), and his "La Florida" map incorporates information from the so-called "De Soto" expedition. Another forty years passed before De Laet issued an American continental map that showed California attached to the mainland (Plate 3). As a director in the Dutch West India Company, he was privy to official information which accounts for the remarkable rendering.

The second phase of Texas cartography was based on details from Spanish reports. The best depiction of southwestern

1

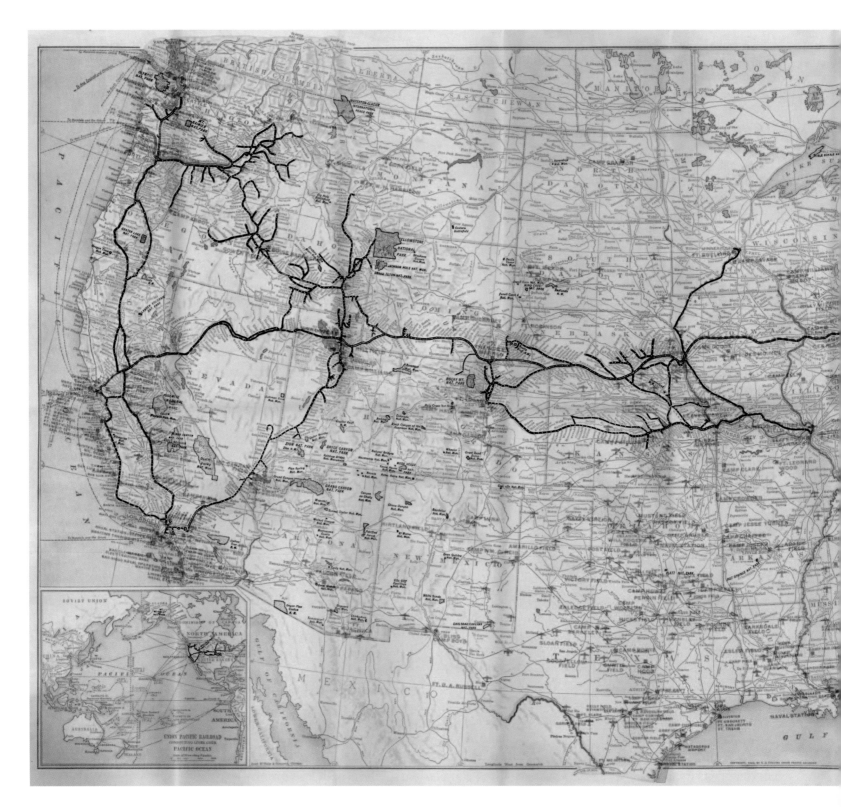

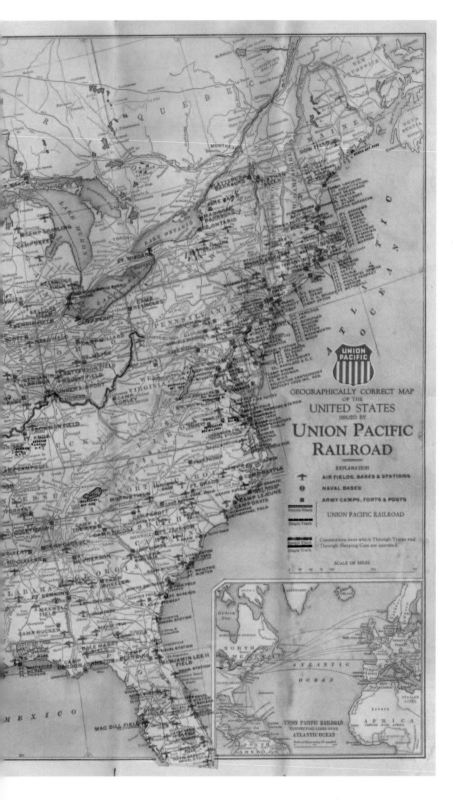

PLATE 61

Military Map of the United States by Rand McNally & Company. Published by Union Pacific Railroad; Chicago, 1942. 17" x 31".

The Union Pacific Map promotes "The Strategic Middle Route" through the country's heartland. The legend lists military camps, bases, and airfields. Texas was permanently changed after the military build-up. The Gulf Coast became the world's largest petrochemical complex. Aircraft factories were in Grand Prairie and Fort Worth. Overall manufacturing increased four times during the war. Texas, representing five percent of the U. S. population, provided seven percent of the armed forces personnel. Over twenty-two thousand Texans died as a result of the war, a third of whom were members of the naval and marine forces. A half million people from other states migrated to Texas to fill the industrial jobs. Cotton and cattle were no longer the greatest economic factors for the state. It was no longer primarily rural. Texas was becoming urban and diversified.

missions was by De Fer in 1705. (Plate 4). The Spanish then made *entradas* to La Junta, in the TransPecos region of Texas, East Texas presidios were established after La Salle's landing at Matagorda, and the subsequent French territorial claim to the Southwest is illustrated by Chatelain's 1719 map (Plate 6).

The British were also interested in the Spanish treasure fleet that sailed annually for Madrid, and Moll's 1715 map of the Galleons Routes (Plate 5) reflects that Anglo interest. English exploration shifted south as pictured in Bowen's 1747 map of New Spain. The insert of an island territory, "the Gallipago" (Plate 7), shows their foothold along the continent. Fifty years later, Kitchen's map described the Southwest as a "Great Space of Land Unknown" (Plate 9) about the time that Spain revitalized its East Texas presence to counter French expansion in the Louisiana Territory. D'Anville's Atlas of 1752 locates the first Texas capital, "Los Adaes" (1721), established to block French incursions from Natchitoches (Plate 8).

The Napoleonic Wars ultimately forced France to sell the Louisiana Territory to the United States, about the time of Poirson's map (Plate 10). After the Louisiana Purchase, as shown in Carey's 1814 Missouri Territory map (Plate 12), Lewis and Clark, and Lieutenant Zebulon Pike explored the West. Pike's East Texas map of 1812 added to the regional geography (Plate 11), whereas Pinkerton's map of the Spanish dominions in North America in 1815 (Plate 13) represents one of many produced that decade.

Mexico declared its independence from Spain in 1821, but much of Mexico remained uncharted. The Brue map of 1825 (Plate 14) describes northern Mexico merely as "Unknown Country," but even more uncertainty surrounded the Big Bend region as displayed in Vandermaelen's map of 1827 (Plate 15).

The third phase of Texas cartography began with the Mexican Federation and extended through Texas statehood. The Tanner map, first published in 1825, best describes the Federal States of Mexico (Plate 16). Due to its popularity, the issue had five editions through 1847, and numerous publishers plagiarized it. Homesteaders demanded more information before entering the frontier, so travel guides were rendered that included maps designed by Hooker (Plate 17), Young (Plate 18), and Lee (Plate 19), as well as many others. The most popular map was by the "Father of Texas," Stephen F. Austin, reissued in a post-Republic edition (Plate 21).

After the Texas Revolution, American and European immigration to Texas exploded. Hunt and Randel used information borrowed from the Texas General Land Office to create a large map of the Republic of Texas (Plate 20). A similar map was issued in Germany by Fleming in 1841 (Plate 23). Sauerlander (Plate 27) and Radefeld (Plate 30) produced two additional Texas maps for the European market. A Vermont atlas produced by Greenleaf in 1842 (Plate 24) provided yet another cartographic version of the new nation. *A History of the Republic of Texas* included a compiled map published by Day and Haghe of London (Plate 22).

As American interest in Texas increased after statehood, David Burr updated his popular map into a new version in 1840 (Plate 29). Regional charts were also needed. Morse and Breese published Gregg's journey to North Texas (Plate 25) in 1844, Williams included an inset of "Texas North of Red River" (Plate 26), and the German Immigration Company added a road map through the Texas Hill Country (Plate 28).

The Mexican War and the California Gold Rush led to more publications, including the 1846 Gilliam crudely illustrated proposed railroad route across Texas to California (Plate 31). The Mexican conflict was well represented by Mitchell's informative "seat of war" map issued in 1847 (Plate 32), and Colton's elephant-size portfolio map pictured all of North America (Plate 3). General Johnston's Topographical Corps of Engineers map of 1850 traces the army's route through West Texas (Plate 37), and Gilman's Treaty Map celebrated the end of the war between the United States and Mexico (Plate 34). Bartlett's Boundary Commission piece (Plate 38) shows the start of a long running dispute over the international boundary line. Gold fever prompted Hutawa's 1849 western map based on General John Charles Frémont's survey (Plate 35). Colonel Carlos Butterfield petitioned Congress in 1859 to improve steamship service to the Pacific Coast as exhibited in his accompanying map (Plate 40). The Civil War prompted a map by Bacon modeled after one issued by Colton in 1862 (Plate 41).

The fourth phase combines government, scientific, and railroad surveys created in the last half of the nineteenth century. Hall's geological map of the Trans-Mississippi was based on army surveys directed by Colonel Emory (Plate 39), and Pressler's western cattle trails, issued by the government, encouraged more army beef contractors for the western forts (Plate 43). A map of telegraph stations published in 1879 (Plate 45) illustrated the frontier's communications network, while the arid Southwest's need for water inspired the Texas artesian well map (Plate 52). The Mexican American boundary commission continued its work through the end of the century, as recorded in Parry's map of 1899 (Plate 54).

Exploitation of the state's natural resources grew with industrialization. A Fort Worth company's oil-well drilling record represents one of the few economic bright spots during the Depression (Plate 57), and technology's future importance to the state's economy is forecast in the Morton Salt Company's Radio Station Map of 1930 (Plate 59).

The concluding maps are divided into four categories: official, military, promotional and transportation. Several publishers, such as Hunt and Randel, John Arrowsmith, James Wilson, and Jacob De Cordova, suggested that their works were officially sanctioned while others, such as Richardson's Almanac Map, became semi-official (Plate 42). Pressler and Langethnann, after long tenures as draftsmen in the General Land Office, gained status by association (Plate 44). The Commissioner of Insurance, Statistics, and History issued a map financed by the state legislature in 1882 (Plate 48), and the state highway system route structure was introduced by the *Dallas Morning News* in 1928 (Plate 58).

Although underwritten by private subscriptions, the "Official Centennial Map" of 1934 celebrated the founding of the Republic (Plate 60). The cartoon rendering of the "Official Texas Brags Map" was so popular that no politician would deny state authorization (Plate 62). The newest release, "Military History Map of Texas," was compiled, drawn, and issued by the General Land Office in 2006 (Plate 64).

Army maps are not as common as the state's complex military history would suggest. A late campaign map was published for the "Operations Against Hostile Indians" in 1886 (Plate 50), even though a longer military episode involved border incursions into Mexico. Rand McNally published a twentieth-century example in 1914, showing army movements, ship stations, and military posts over a previously issued Mexican map (Plate 55). Four years later, during World War I, the mapmaker adopted a standard edition for the Rock Island Railroad Lines to locate National Guard camps, cantonments, officer training schools and aviation sites during World War I (Plate 56). The Union Pacific Railroad did likewise in 1942 when it overprinted another map with its "Strategic Middle Route" that listed World War II camps, bases and air fields (Plate 61).

Promotional maps became standard fare for mapmakers from the time of the Republic. For example, the proprietors of the German Immigration Company provided plans of four central Texas cities to lure homesteaders (Plate 36). Railroad companies handed out timetables, guide books, and sales-brochure maps promoting company-owned acreage (Plate 46), while other railroads transferred land to real estate holding companies, such as the New York and Texas Land Company, Ltd., which held three million acres in the Panhandle (Plate 47). Late-nineteenth-century developers proposed new commercial and recreational centers such as one in Velasco (Plate 53), but some communities, solely out of civic pride, commissioned bird's-eye views such as the one for the railroad center at Denison, Texas (Plate 49).

The final map in this collection was an advertisement marketing air travel to Texas: a 1952 American Air Lines magazine ad campaign surrounded a map of the United States with boarding destinations represented by ticket stubs to emphasize the ease of travel; directions and landmarks were no longer necessary according to this new age *portolano* (Plate 63).

When Texas joined the Union in 1845, the state was twice its present size, and in 1850 the state received $10 million for relinquishing claim to parts of Kansas, New Mexico and Colorado, a total of almost 266,000 square miles. Today Texas consists of about 262,840 sections or 168 million acres. Its eastern border, near the Sabine River, was established in 1819 by the Adams-Onis Treaty, and the border with Mexico was not finalized until the twentieth century. For a half century, Texas was more than the United States could either survey or protect. The Department of Texas was even a separate army command for almost thirty years.

The vastness of the state frustrated the United States Army, Texas Rangers and General Land Office surveyors for decades, leaving most of the land uncharted. On March 20, 1848, the second state legislature passed an act that required each county to hire a mapmaker to record its region. A "suitable draftsman" was to draw two correct copies of General Land Office maps for use in each district, with one to be copied on "good tracing paper" for ease of reproduction. As late as 1904, a United States Government Printing Office publication, *A Gazetteer of Texas*, noted that "the subdivision of Texas lands has not resulted in the production of maps of much value. The coast line has, however, been mapped by the United State Coast and Geodetic Survey, and since 1884 large areas in the central and western portions of the state have been mapped by the Geological Survey . . . more than one-fifth of the area of the State." The federal government left the mapping of Texas to the General Land Office, but surprisingly the cartographic record was more complete than the federal publication indicated, even if many maps were privately produced. That is why, even today, publicly accessible map collections remain essential to scholars and researchers. *Going to Texas: Five Centuries of*

Texas Maps aims to take the beauty of maps and the history that is visually presented by them beyond scholars and researchers to all those who want to know more about the discovery, settlement, and development of the great State of Texas.

SUGGESTED ADDITIONAL READINGS

Arbingast, Stanley, et al. *Atlas of Texas* (5th ed.). Austin: University of Texas, 1976.

Bryan, James P., and Walter K. Hanak. *Texas in Maps.* Austin: University of Texas Press, 1961.

Martin, James C., and Robert Sidney Martin. *Maps of Texas and the Southwest*, 1513-1900. Albuquerque: University of New Mexico Press, 1984.

_____. *Contours of Discovery: Printed Maps Delineating the Texas and Southwest Chapters in the Cartographic History of Texas.* Austin: Texas State Historical Association, 1982.

Pool, William C. *A Historical Atlas of Texas.* Austin: The Encino Press, 1975.

Spanish Explorations and Expeditions into Texas, 1519-1783

Donald A. Chipman

After the initial voyage of discovery by Christopher Columbus, the Spanish empire in America began amid the major islands of the Caribbean Sea. Española, or Santo Domingo, as it is better known, was permanently settled by Columbus in 1493. By 1508 Puerto Rico and Jamaica had fallen, respectively, under the control of Juan Ponce de León and Juan de Esquivel. And in 1511 Diego de Velázquez initiated the conquest of Cuba, the largest island in the Antillean chain. From Puerto Rico in 1513, Ponce de León first touched Florida, which he thought to be an island—a geographic error that would persist in the mind of Spaniards for the next six years.

In early 1519 Fernando Cortés left Cuba and established the initial Spanish settlement on the coast of Mexico to the north of modern-day Veracruz. At the very same time, Francisco de Garay, Esquivel's successor as governor of Jamaica, launched a Gulf Coast exploration that brought Spaniards along the Texas coast.

Garay's choice as captain of this sea expedition was Alonso Alvarez de Pineda. On board four vessels was a crew of 270 men. Pineda set sail from Jamaica and proceeded through the Yucatan Channel to the western tip of south Florida. From there he attempted to sail northeastward, but strong head winds forced him to turn about. He then ran the Gulf Coast from the Florida Keys to Cortés' settlement named Villa Rica de la Vera Cruz. En route, it is likely that one of Pineda's pilots drew the earliest known map that depicts the Texas coast. This 1519 map also establishes that Florida is a peninsula, although it is severely truncated and placed too close to Cuba.

In the course of his voyage, Pineda noted the discharge of several large rivers into the Gulf but named only two of them. He recorded the Mississippi River as the Río del Espíritu Santo, because it was discovered on the day of the feast of Espíritu Santo (June 2, in 1519). Significantly, the second named stream was the Río Pánuco, which enters the Gulf at present-day Tampico, Tamaulipas.

Pineda continued to navigate the Gulf Coast to an anchorage off Villa Rica de la Vera Cruz, where he arrived just hours after Cortés had departed for the conquest of the Aztec Empire. Unwilling to leave his base camp in the proximity of a potential rival, Cortés returned to the coast and through a series of ruses and threats persuaded Pineda to lift anchor and return to the north.

Rebuffed at Villa Rica, Pineda then sailed up a large river where he spent about forty days. The location of that river is extremely important to the history of exploration in the early sixteenth century. Carlos E. Castañeda, renowned historian and author of *Our Catholic Heritage in Texas* (7 vols.), identified it as the Río Grande, "soon to be known as the Río de las Palmas."

To support his contention, Castañeda cited the work of a sixteenth-century historian, López de Gómara, who stated the distance between the Río Pánuco and Río de las Palmas (i.e., the Río Grande, according to Castañeda) as slightly more than thirty leagues or about ninety miles. This statement in itself represents an astonishing oversight by Castañeda. Along the coast from Tampico to the mouth of the Río Grande, the distance is approximately 250 miles; by road it is well in excess of 300 miles. Castañeda also felt obliged to refute evidence flowing from an expedition along the coast of Mexico led by Alonso de León, Sr., in the 1650s. De León recorded the mouth of the Río de las Palmas at twenty-four degrees north latitude, which is only minutes off the present-day Río Soto la Marina where it enters the Gulf at La Pesca. Undaunted, Castañeda insisted that De León made a mistake in reading in his astrolabe and that he should have reported twenty-six degrees north latitude. Several early maps

PLATE 1

Nueva Hispania Tabula Nova by [Jacopo] Gastaldi. Published in Atlas,
La Geografia: Venice, 1548. 5" x 6.75".

*This first map for the general public is devoted entirely to New Spain and
the Gulf of Mexico Coast. Yucatan is shown as an island. Texas is depicted
as mountainous. This map of the American Southwest was used as a
model for Ruscelli's version thirteen years later. This is considered the first
atlas to use copperplates for print transfer and the first "pocket atlas."*

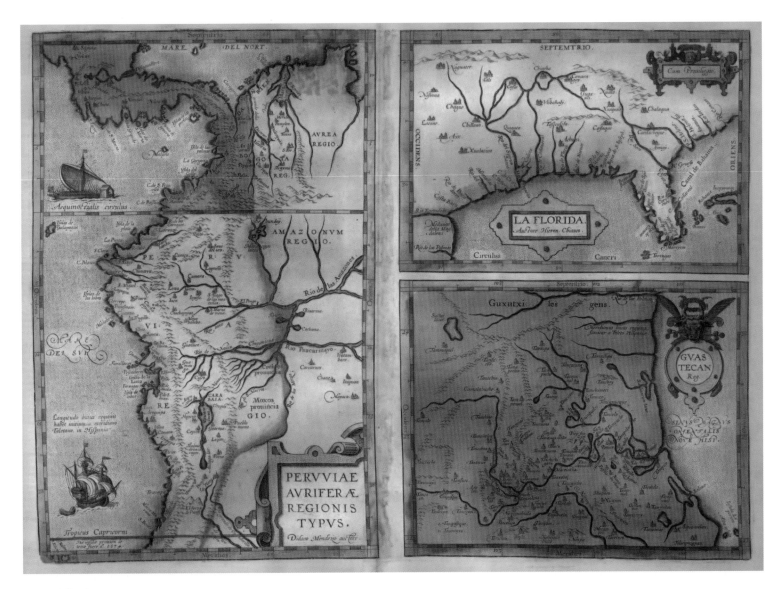

PLATE 2

Peruviae Auriferae Regionis Typus; La Florida; Guastecan by
Abraham Hieronymo [Gerónimo] Chaves Ortelius and Didaco Mendezio
(Diego Mendez). Published in *Theatrum Orbis Terrarum. Additaemntum
III*; Antwerp, 1584. 14" x 18".

*This is a copper engraving of the first detailed map of the Spanish coast-
line along future Texas. It is the first published map of Florida based on
DeSoto's charts. The South American image locates gold sources in Peru
and silver in Bolivia. Latin text on verso. Wytfliet's "Florida et Apalche"
plagiarized Ortelius', "La Florida" (1597). The Mississippi River,
although distorted and titled "R. De S. Spirtu," reasonably traces the
course of the river.*

PLATE 3

Americæ sive Indiae Occidentalis, Tabula Generalis by Joannes De Laet. Published by Hessel Gerritaz in *L'Histoire du Nouveau Monde ou description des Indes accidentales* or Dutch Edition: *Beschrijvinghe van West-Indien*; Leiden, 1630. 11" x 14".

Although sparse of the usual embellishments for period, the cartography is remarkably accurate for the time. De Laet's latitudes are close to actual. He correctly shows California attached to the mainland and not an island as in other contemporary atlases. He obviously knew of Spanish cartographer Antonio's de Herrera's "Description de las Yndies del Norte" (Madrid 1601). De Laet, as a director and shareholder in the Dutch West India Company, had access to the company's latest American intelligence. The company's chief cartographer, Hessel Gerritaz, apprenticed under William Bleu. The map's style is a stark departure from the heavier engraving used by Mercator and Ortelius. Outline coloring is not original.

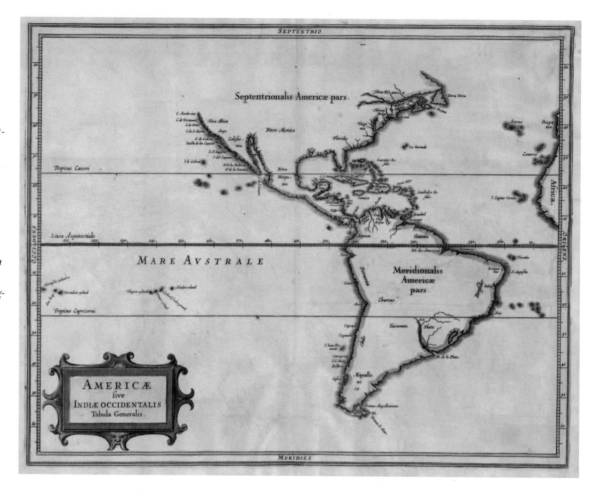

position the Río de las Palmas at precisely the same location as the Río Soto la Marina on a modern map of Tamaulipas.

In support of contentions by Castañeda and more recent writers that the Río Grande was the Río de las Palmas of colonial times and that it was the river entered by Pineda in 1519 where he remained for more than a month, a red herring surfaced in 1974. A naval reserve unit based in Harlingen, Texas, on a weekend dig at the Boca Chica site near the mouth of the Río Grande, unearthed a carved stone tablet. Its inscription translated into English reads: "Here. Alonso Alvarez de Piñeda. Captain. 1519. With 270 men and 4 ships." On one side of the stone are scratched the words: "Colony of Garay."

Donald Chipman played a significant role in exposing the so-called "Piñeda Stone" as a hoax. Block letters on the stone are serif, definitely not common in sixteenth-century Spain; the letters "CTN" for captain are not recognized by Spanish paleographers as an acceptable abbreviation for *capitán*; all of the numbers on the tablet are in Arabic, and while Arabic numbers were becoming somewhat generalized in Castilian by that time, Roman numbers were almost exclusively used in monumental epigraphy. Two other clues are even more damaging to claims of authenticity for this stone. Spanish genealogists insist that "Piñeda" (with a tilde) does not exist as a surname in their country; and most important of all, the "7" in 270 has a slashed stem, a common rendering today in European countries, but the slashed-stem "7" did not appear until the latter part of the nineteenth century.

Finally, evidence suggests that Pineda actually sailed up the Río Pánuco in 1519, not the Río de las Palmas *or* the Río

PLATE 6

Carte Contenant le Royaume du Mexique et la Florida by Henri Abraham
Chatelain. Published in *Atlas Historique*; Amsterdam, 1719. 16.5" x 21".

*This map is a copy of Guillaume Delisle's 1703 "Carte du Mexique et
de la Florida" (Lowery 256; Martin & Martin14) and remained the
standard for a half century. It has expanded remarks and more colorful
decorations and incorporates information from reports filed by survivors
of La Salle's expedition into Texas. The map accurately traces the
Mississippi River for the first time.*

PLATE 4

Carte de California et du Nouveau Mexique by Nicolas deFer. Published in *L'Atlas Curieux*, Paris, 1705. Lithograph by Charles Inselin. 8.75" x 13.5".

Copper engraved, this is the first published chart of Jesuit missionary Father Eusebio Francisco Kino's exploration of the Southwest in 1694. He reduced it to a manuscript map in 1695 and sent it to the Mexican Viceroy. California is still shown as an island. He connects it to the mainland in later charts, which was considered a rediscovery of the California peninsula. Places and missions are keyed to numbers, including several in Big Bend region of Texas. According to Tooley, deFer's is a "highly important map." This profound map illustrates Father Kino's marked influence on other maps for over a century.

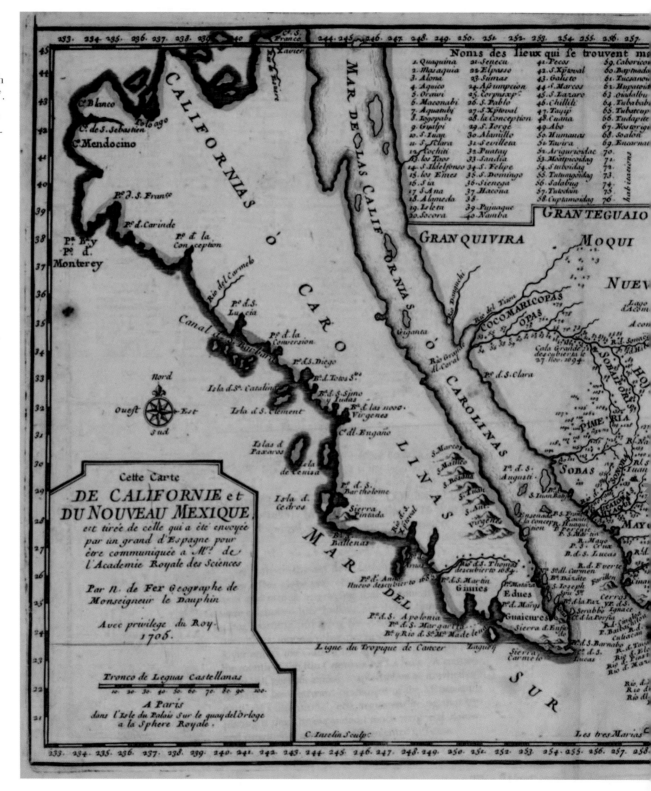

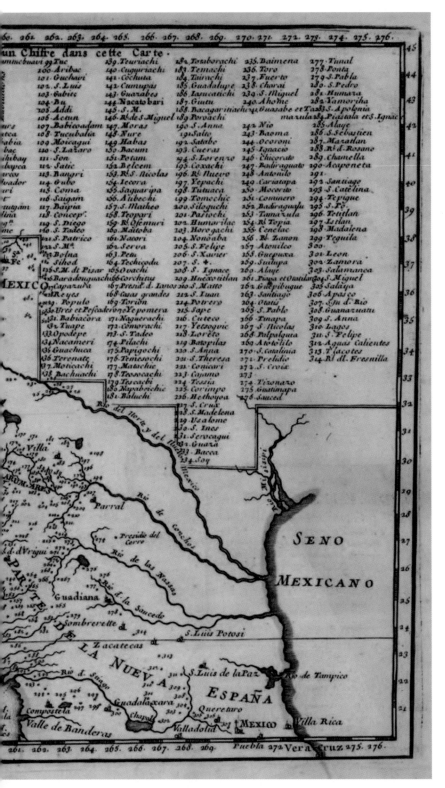

Grande. In the following year, Pineda and another Spanish captain returned to this river, whose banks were populated by Huastec Indians. Shortly after their arrival, the Indians revolted and killed some forty Spaniards, including Pineda. Three years later, Spaniards led by Cortés conquered this region of Mexico. There they found skins of their deceased countrymen, tanned like fine leather with their beards still intact, which they used to identify old acquaintances in the islands. And, during this conquest, Spaniards constructed boats "made from the wood of old vessels that had belonged to the captain sent by Garay, who had been killed."

In any exhibit or publication devoted to explorers and expeditions into Texas, it is important to establish that the first Spaniards on the soil of the future Lone Star State did not arrive until November 1528. Those men were survivors from a sizable expedition that had landed near Tampa Bay, Florida, in that same year.

The circumstances that brought Alvar Nuñez Cabeza de Vaca to Texas, along with nearly 250 fellow Spaniards, are too well known to dwell on. What is important is the likely route of the only four survivors of this group across Texas and into Mexico. It seems apparent that five barges launched from the area of modern-day Pensacola, Florida, each containing approximately fifty men, arrived on the Texas coast to the west of Galveston Island. In just a few months, almost all of these Spaniards were dead. Some drowned; others died of hunger, exposure, or snake bite; still others succumbed to Indian attacks or to homicides inflicted by fellow Spaniards, followed by incidences of cannibalism.

By the winter of 1532–1533, only four of the original number had survived. They were Cabeza de Vaca, two other Spaniards, and an African slave. At that time all were slaves of Texas Indians who treated them badly. In mid-September 1534, these men, commonly called the Four Ragged Castaways, escaped from their Indian masters and fled toward Mexico.

Their route across Texas has spawned nearly as many interpretations and projections as there are writers on the subject. Until recent years, the proposed trekking of these men may be broadly labeled Trans-Texas, beginning near Galveston Island and ending near Presidio or El Paso. Proponents of this pathway to Mexico by the four men have been heavily influenced by Texas nationalism and localism, with the apparent intent of placing the wanderers in as many Texas counties as possible. For example, the late Dan Kilgore of Corpus Christi, a former president of the

Texas State Historical Association, noted that almost half of Texas' 254 counties have claimed that Cabeza de Vaca had trod their soil. Some community enthusiasts have even argued over which side of their main street was traversed by a man often called "The Great Pedestrian."

Current scholarship points to an inner coastal route of the castaways, starting at Matagorda Bay, crossing the Nueces River near modern-day Corpus Christi and then following a south-westerly course to a crossing of the Río Grande near present-day Roma, Texas. This route interpretation is supported by flora and fauna described by Cabeza de Vaca in the first literature on Texas, by distance of travel set forth in his narrative, by the cast-aways' intentions of reaching Mexico as quickly as possible, and by their determination to avoid hostile Indians along the Texas coast who had murdered many of their former comrades.

After crossing into Mexico, Cabeza de Vaca and his companions eventually reached Mexico City in July 1536. In all, these four men traveled an estimated 2,500 miles, most of it on foot. They told and retold stories of their experiences to eager audiences. With the exception of truth, their stories of rich lands *"más allá"* lost little in retelling.

Soon powerful men capable of mounting a follow-up expedition into the north country vied for the opportunity of finding the legendary Seven Cities of Cíbola. The romantic notion of cities with streets paved in gold, or El Dorado, would soon propel Spaniards into New Mexico and beyond to Texas.

Between 1540 and 1542, Francisco Vázquez de Coronado traveled across parts of New Mexico, Texas, Oklahoma, and

PLATE 5

***A Map of the West Indies or the Islands of the Americans . . . .
Explaining What Belongs to Spain, England, France, Holland & C . . .
Ye Several Tracts Made by Ye Galeon's and Flota from Place to Place***
by Herman Moll. Published by Thomas Bowles and John Bowles;
London, 1715. 23" x 40".

*This map shows the Gulf Coast and a widened Florida along with de-
tailed trade routes from Central America and Mexico across the Atlantic
Ocean. The birds-eye view of Mexico City is classic. Insets show harbors
at Vera Cruz; Havana; Porto Bello; St. Augustine; and Cartagena.
The map clearly reflects English naval interest in Spanish treasure
ships through Gulf of Mexico and up the south Atlantic coastline along
Georgia's Sea Islands. Louisiana extends to the Rio Grande River.
New Orleans, founded in 1718, is absent. Florida is shown as English
while Texas is shown as French.*

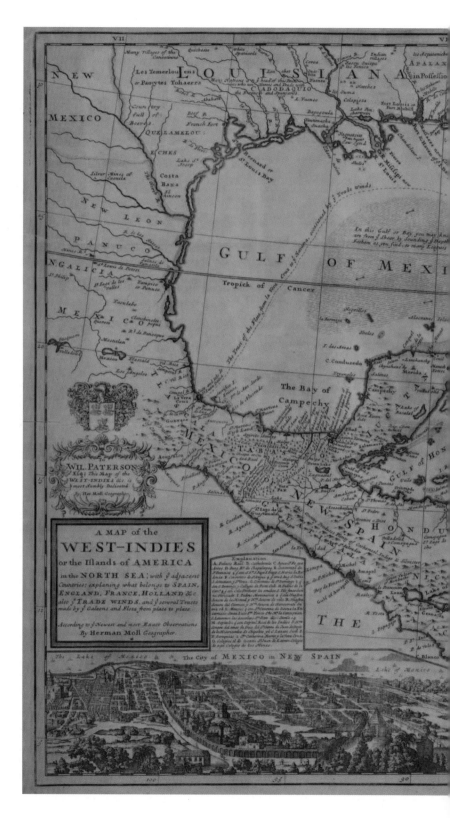

PLATE 7
**A New & Accurate Map of Mexico or New Spain . . . California, New
Mexico** by Emanuel Bowen. Published by W. Innys in *A Complete Atlas*;
London, 1747. 15" x 18".

*This map includes an inset of "The Gallipago Islands Discovered &
Described by Capt. Cowley in 1684." The cartouche shows English
explorers with conquered natives under foot. The map illustrates
growing British interest and influence in the region.*

Kansas in search of gold. This expedition left the town of Culi-acán in western Mexico with about four hundred adults and young men. It was accompanied by hundreds of Indians and a few African-born slaves who carried supplies and equipment. After a disappointing failure to find exploitable wealth among the Pueblos of New Mexico, Coronado set out in spring 1541 for the plains of eastern New Mexico and the Texas Panhandle.

While crossing part of the Texas Panhandle, Coronado and his army camped near a large canyon, probably Tule on Prairie Dog Town Creek. There Spaniards encountered the harshness of West Texas weather for the first time. A thunderstorm with high winds and dark clouds soon brought hailstones "as big as bowls, or larger," which fell "as thick as raindrops." The hailstones dented Spanish helmets and forced soldiers to take cover under their shields.

After this violent storm, Coronado sent the bulk of his army back to the Río Grande in New Mexico. He then force-marched a small contingent of men toward a promised land of wealth called Gran Quivira along the Arkansas River in Kansas. Once there he found only Wichita Indians who lived in grass huts with nearby fields of corn.

Forced to admit his failure to find gold, Coronado followed a more direct route back to New Mexico that paralleled the course of the Arkansas River and eventually crossed the extreme north-west corner of the Texas Panhandle. Back in New Mexico, Coronado wrote a letter to his king in which he described lands that no other Spaniards, other than those in his army, had ever seen. The Llano Estacado (stockaded not staked plains) and the Great Plains were described by him in these words: "[I] reached some plains, so vast that I did not find their limit anywhere that I went, although I traveled over them for more than 300 leagues [about 900 miles]." He also reported seeing so many bison that it was "impossible to number them."

In spring 1542, Coronado returned without incident to Mexico by way of Culiacán and then on to Mexico City. His expe-dition was the second to probe the periphery of Texas. By sheer coincidence, a third expedition was on its way to Texas from the east. These men were remnants of an expedition led by Hernando de Soto, which had landed in Florida in 1539 (Plate 1).

Soto's huge army covered hundreds of miles from May 1539 to May 1542, primarily in the present states of Florida, Georgia, Alabama, Louisiana, and Arkansas. In all their wander-ings and deadly attacks on Indians, a small chest of inferior qual-ity fresh-water pearls was the extent of plunder that fell into their hands, and even this booty was lost during a fight with Indians in Alabama (Plate 2).

PLATE 8

Map of Louisiana

from *D'Anville's Atlas* by Jean Baptiste Bourguignon d'Anville. Published by Jno. Harrison; London, 1791. Lithograph by Bowen; drafting by Haywood. 20.5" x 36".

This republication of D'Anville's map of 1756 was done for Harri-son's Atlas. It clearly identifies the location of "Los Adaes" (Adaxes), earliest capital of Spanish Texas, established under the patronage of the Duke of Orleans by D'Anville.

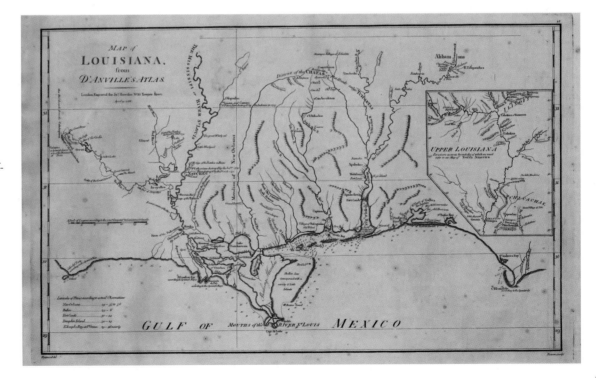

PLATE 9

Mexico or New Spain in which the motions of Cortés may be Traced
by Thomas Kitchin. Published by W. Strahan and T. Cadell in William
Robertson's *History of America*; London, 1777. 11" x 15".

*The inset is of the environs of Mexico City. The vast land region north of
Texas is described simply as "Great Space of Land Unknown." This is
an English map issued during American Revolution. Cortés' exploration
is noted. Kitchin continues to use some names given to islands by Father
Kino in his map of 1710.*

A raging fever claimed the life of Soto on May 21, 1542. His successor, Luis de Moscoso Alvarado, then led remnants of the army westward from the Mississippi River, seeking an overland route to Mexico. Evidence suggests that the farthest extent of Moscoso's march into Texas was the Brazos River in the region of College Station. However, some Texas nationalists have placed his crossing of the Brazos in Young County, about eighty miles west of Fort Worth.

Moscoso, recognizing the vastness of Texas, retraced his route to the Mississippi River and arrived there in late 1542. There he decided to build seven small craft and descend the Mississippi with 322 men. From the Gulf Coast, he navigated the entire shoreline of Texas and beyond to the Río Pánuco. Remarkably, Moscoso lost only ten men.

The failure of both Coronado and de Soto to find wealth in lands north of Mexico was so discouraging that Spaniards did not again enter the East Texas area for the next 143 years. Even then it was not with the hope of finding wealth; rather it was to counter French designs on the future Lone Star State (Plate 3).

In early 1685 René Robert Cavelier, Sieur de la Salle, apparently by mistake, landed on the Texas coast and founded a French colony on Garcitas Creek near Matagorda Bay. News of the French intrusion into lands claimed by Spain reached Mexico in the following year. Starting in 1686 and ending in 1689, the Spanish dispatched six overland expeditions and five undertakings by sea with the intent of finding and destroying the French outpost in Texas (Plate 6).

By the time Alonso de León, Jr., found the site of the French colony in April 1689 on his fourth overland attempt, it had been destroyed by an Indian attack in late December of the previous year. Although La Salle had been murdered by his own men in East Texas and the French threat to Texas temporarily thwarted by disease, internal dissension, and Indian attacks, Spain recognized the necessity of establishing a foothold in Texas. This led to the founding in 1689 of San Francisco de los Tejas, the first Spanish mission in East Texas. The location of this mission, according to recent research, was on San Pedro Creek just east of present Augusta, Texas.

Resupply of the mission depended on expeditions sent from Mexico, especially one led by Domingo Terán de los Ríos in late 1691. Terán, the first Spanish governor of Texas, explored Texas north of the mission as far as Kadohadocho villages on the Red River. The expedition was plagued by freezing rain and more than a foot of snow. Terán's nightmarish experiences in Texas fueled negative impressions of the province that had a lasting effect. Speaking of the province in his diary, he remarked that "no rational person has ever seen a worse one."

After Terán's departure from Texas, conditions at San Francisco de los Tejas, exacerbated by the spread of European-introduced diseases among the Tejas, had deteriorated to near flash point by fall 1693. Fearing for their lives, the padres buried large objects such as bells and cannons, packed sacred items, and then burned the mission before retreating to the safety of Coahuila.

Given Terán's baleful assessment of East Texas, the failure of the mission there, and the apparent end of French incursions, it is hardly surprising that not a single Spaniard remained in the region from 1694 to 1715. That, however, would soon change (Plate 4).

After the death of La Salle, French interests in the lower Mississippi and territory to the west of it remained alive at the court of Louis XIV. The king's minister of marine and colonies and Count of Pontchartrain was Louis Phélypeaux. Count Pontchartrain was quick to perceive the advantage of establishing a permanent French colony in the region. It would secure access to the Gulf from French Canada, it would protect against English and Spanish designs in the area, and it would place France in a position to attack rich Spanish silver mines in northern Mexico (Plate 5).

Chosen to lead the French colonization effort of 1699 was Pierre Le Moyne, Sieur d'Iberville. Accompanying Iberville were four of his brothers, the most important of whom was Jean-Baptiste Le Moyne, Sieur Bienville, the future "Father of Louisiana." Far more important to Texas, however, was Iberville's Canadian-born cousin by marriage, Louis Juchereau de St. Denis.

St. Denis was lured toward Texas in 1713, at least in part, by the French desire to make Louisiana a profitable proprietary colony. En route, the French adventurer stored goods at the future site of Natchitoches before continuing on to a Spanish settlement just south of the Río Grande. His arrival there in July 1714 touched off a renewed determination by Spanish officials in Mexico City to secure East Texas against French incursions.

However, it was April 1716 before a sizable colonizing effort led by Domingo Ramón crossed the Río Grande into Texas. Counting priests, soldiers, and civilians, the contingent numbered seventy-five persons. Significantly, seven of the soldiers were married and brought along their families—making their wives the first Spanish women to enter East Texas.

The Domingo Ramón expedition established a Spanish presence in Texas that would last from 1716 to 1821. Texas would nevertheless witness a shifting scene of abandoned and restored missions and presidios. Indeed, the six missions in East Texas set up by Franciscans in 1716–1717 would be abandoned in 1719 because of panic over a minor war in Europe that pitted France against Spain. This conflict spread to the border of French Louisiana and Spanish Texas, resulting in East Texas once again being abandoned by Spain (Plate 7).

In 1721 Spain reclaimed East Texas, thanks to efforts of the Marqués de Aguayo, a rich Spaniard in northern Mexico. It is hard to overstate the importance of the Aguayo expedition. It established the cattle industry in East Texas; it reestablished the six missions there that had been abandoned in 1719; it constructed two new presidios; and it increased the military from sixty or seventy to 268. In the words of one historian, "[The Aguayo expedition] was perhaps the most ably executed of all . . . that entered Texas, and in results it was doubtlessly the most important" (Plate 8).

From 1721 to the 1780s, Spanish control over Texas was essentially unchallenged by any foreign power. In large measure this may be attributed to a steadily progressing rapprochement between Spain and France, both ruled by Bourbon dynasties,

that culminated with the Bourbon Family Compact of 1761. Texas was further strengthened in the early 1770s by a royal order known as the "New Regulations for Presidios." This reorganization, among other less important provisions, led to moving the capital of Texas from Los Adaes in East Texas to San Antonio. By the 1780s, however, the Spanish empire north of Mexico from Florida to Texas would come under increasing pressure from the newly independent United States of America, and interest grew about the "Great Space of Land Unknown" (Plate 9).

SUGGESTED READINGS

Castañeda, Carlos E. *Our Catholic Heritage in Texas, 1519-1936*. 7 vols. Austin: Von Boeckmann-Jones, 1936-1958.

Chipman, Donald E. *Spanish Texas, 1519-1821*. Austin: University of Texas Press, 1992.

Foster, William C. *Spanish Expeditions into Texas, 1689-1768*. Austin: University of Texas Press, 1995.

Tyler, Ron, et al, eds. *The New Handbook of Texas*. 6 vols. Austin: Texas State Historical Association, 1996.

Weddle, Robert S. *The French Thorn: Rival Explorers in the Spanish Sea, 1763-1772*. College Station: Texas A&M University Press, 1982.

DEFINING THE NEXUS OF EMPIRE

*The Louisiana Purchase
and Texas Borderlands, 1803-1821*

GENE ALLEN SMITH

———◆———

Frenchman Arsène Lacarrière Latour fought in Haiti in 1803 against the black slave uprising that engulfed the island and destroyed French rule, practiced as an engineer and architect in New Orleans during the first decade of the nineteenth century, and in January 1815 served as Andrew Jackson's principal engineer at the Battle of New Orleans, designing defenses that permitted an American victory at Chalmette. Once quiet returned to the Gulf region, Latour resumed his life of adventure, selling his services to the highest bidder.

During the early summer of 1816 Latour traveled to Philadelphia where he met separately with Jean Lafitte, Spanish Minister Luis de Onís, and American Secretary of State James Monroe. After those meetings, Latour led a special expedition into the "Interior Provinces" of the Southwest for the Spanish government. From June to November 1816 Latour and Jean Lafitte traveled extensively throughout the Southwest, surveying and preparing maps. Latour's exhaustive report to the Spanish government that accompanied the maps revealed that the expedition had visited the headwaters of the Red, Sabine, Arkansas, and Colorado rivers. But more importantly, the report warned that should the Spanish fail to populate and defend the region, "the time will come, and unfortunately is not . . . far off, when the Americans . . . will pour in myriads into Mexico." Although the report received careful consideration and produced extensive Spanish commentary, officials could do little to counter the threat that Americans posed to the Texas borderlands.

Spain had consolidated her rule over Texas and the western Gulf Coast with the end of the Seven Years War (French and Indian War)—a conflict that began as an Anglo-French contest over control of the Ohio River Valley but quickly spread into a world-wide war fought in Europe, Asia, and in North America.

Once the struggle began, the intricate system of alliances intended to maintain a balance of power in Europe took effect. Britain's ally Prussia used its military power to keep the French army occupied, while British naval power isolated French colonial settlements abroad, ultimately forcing France to seek allies willing to join the conflict on its behalf. Spain entered the war in 1761 on the French side hoping to regain control over Gibraltar, but British forces instead retained control over the rocky outcropping and ultimately captured the two valuable Spanish colonies of Cuba and the Philippines. During the Paris negotiations that ended the war, Spain begrudgingly traded control over the colonies of East and West Florida to Britain for the return of Cuba and the Philippines. The treaty also transferred France's North American possessions east of the Mississippi River and in India to the British. The French relinquished Louisiana to Spain for her support in the latter years of the war. Ultimately, the treaty dismantled France's colonial empire and reinvigorated a British North American empire that thereafter stretched from Hudson Bay in the north to the Gulf of Mexico in the south and from the Atlantic Ocean west to the Mississippi River. And though Spain did not secure control over Gibraltar, gaining Louisiana and the two Floridas at the end of the American War for Independence twenty years later (1783) gave the Spanish complete control over the entire north shore of the Gulf of Mexico. Literally, with the stroke of a pen at two Paris peace conferences, the province of Texas evolved from a frontier borderland to an interior province, seemingly far removed from potential conflict (Plate 10).

New threats to the Spanish Gulf Coast colonies began with the emergence of the United States as a nation. Throughout the 1780s and 1790s, Americans looked at the Mississippi River and

21

the Florida peninsula as obstacles to future American growth and expansion. In 1785-1786, during the Confederation period, American Secretary of Foreign Affairs John Jay negotiated a treaty with Spain's minister to the United States, Don Diego de Gardoqui, proposing to close the Mississippi River for at least twenty-five years in return for favorable commercial concessions. Yet once Jay learned of the overwhelming American opposition to the agreement, his better judgment prevailed, and he did not submit the treaty to the Confederation Congress for approval. Although the river remained open, Americans still lacked the right to trade in Spanish-controlled territory.

While the Gulf South—from Texas to Florida—remained under Spanish control, during the 1780s, Thomas Jefferson believed that those lands could not be in better hands. Spain was too feeble to hold them, and Jefferson professed that in time Americans would take the region from Spain piece by piece. Negotiation of the 1795 Pinckney Treaty (Treaty of San Lorenzo), which provided the United States with navigation of the Mississippi River through Spanish territory and the right of deposit at New Orleans, appeared to delay Jefferson's prediction by removing the major obstacle to harmonious Spanish-American relations. Yet that harmony suddenly disappeared when Spain and France secretly agreed to the Treaty of San Ildefonso in October 1800. The agreement, negotiated according to Napoleon Bonaparte's wishes, transferred Louisiana to France and appeared to make the colony a food source for a potential western French empire based in the Caribbean (Plate 13).

When President Jefferson learned of the transfer in 1801, he concluded that a French presence on the Mississippi River and Gulf Coast would result in a war between the two countries. Working through official diplomatic channels and through unofficial American agents in France, Jefferson gained additional information convincing him in early 1803 to send James Monroe to France to assist Robert R. Livingston in gaining control of the

PLATE 10

Carte des Deux Florides et de la Louisiane Inférieure by Jean Baptiste Poirson. Published by C. C. Robin and F. Buisson in *Les Voyages of Mr. Robin*; Paris, 1807. 17.5" x 27.5".

The Texas provincial cities of St Antoine de Bexar and Nacogdoches are located. This map, showing the Gulf Coast according to French cartography, was copied from older Spanish sources, with updates based on Jose Antonio de Evia's Gulf Coast charts published in Madrid in 1799. Includes C. C. Robin's Voyage to Louisiana and Florida between 1802 and 1806.

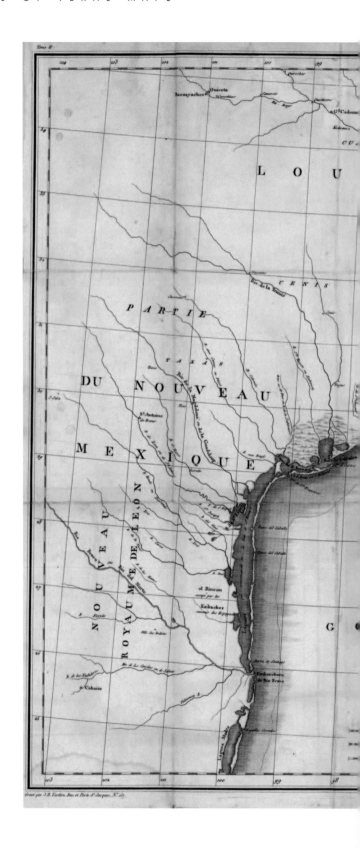

PLATE 13

Spanish Dominions in North America Northern Part by John Pinkerton. Published by T. Cadell & W. Davies in *A Modern Atlas* ; London, 1815. Lithograph by L. Hebert. 20"x 28".

Pinkerton combined information from both Humboldt's and Pike's maps to create this presentation, making it arguably the most current map of the northern province of Mexico for the period. It shows all of Texas north of the Platte River, plus much of western Louisiana and Arkansas Territories. All was taken directly from Pike's reports.

lower Mississippi River, even if it meant purchasing the Isle of Orleans and possibly the Florida peninsula. On April 11, 1803, just a few days before Monroe arrived in Paris, French finance minister François de Barbé-Marbois offered Livingston a chance to purchase all of Louisiana west of the Mississippi River for $15 million. Livingston and the newly-arrived Monroe agreed to the deal on May 2, 1803, doubling the size of the United States and moving the area of potential American-Spanish conflict from the Mississippi River west to the Sabine River and Texas.

The Spanish government protested the illegality of the Louisiana Purchase, arguing that Bonaparte had no right to sell

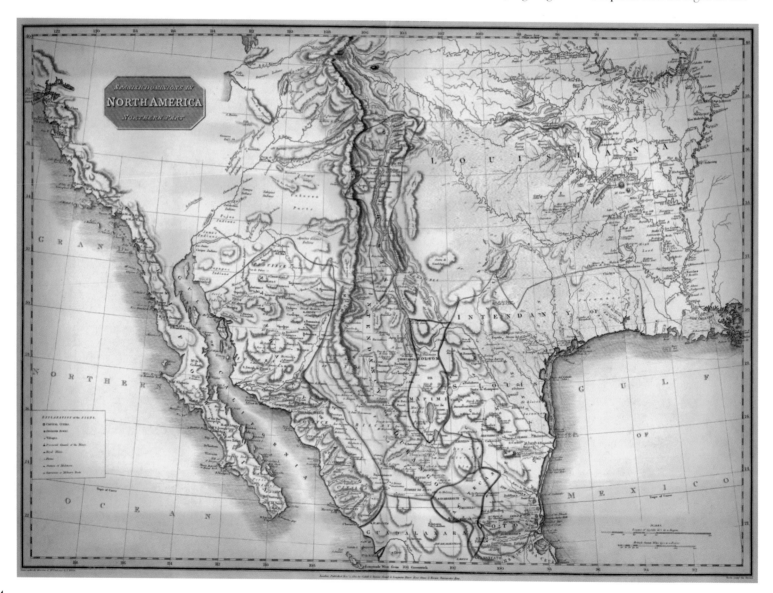

the territory. Bonaparte, the Spanish maintained, had not fulfilled the conditions of the Treaty of San Ildefonso, and thus the territory should have reverted to Spain rather than being sold to a third party. Being unable to rescind the deal, Spanish officials argued that such a purchase could only consist of the west bank of the Mississippi River and the city of New Orleans. French diplomats had purposely left vague the definition of the western boundary of the Purchase, suggesting that the Americans make the most of their deal.

The Americans acted quickly, contending that the purchase stretched southwest to the Rio Grande River and west to the Rocky Mountains—an assertion that would have stripped the Spanish of Texas and New Mexico.

The process of transforming Texas from an interior province to a defensive borderland adjacent to the United States revealed that Spain had done very little to strengthen its position in this region of North America. After 1783, Texas no longer formed the eastern boundary of New Spain. Instead, the boundary stretched across Louisiana and the Florida Gulf Coast. The sparsely settled and poorly defended region of East Texas, which had but a single settlement at Nacogdoches, thereafter became part of the Internal Provinces governed by the Viceroyalty of Mexico; the settlement of Natchitoches, Louisiana, just east across the Sabine River was within the jurisdiction of the captaincy general of Cuba. The invisible provincial boundary between these two bureaucratic entities served only to stagnate the economic development of East Texas. For example, the Spanish bureaucracy dictated that the people of Nacogdoches had to buy expensive and highly taxed goods transported overland from Vera Cruz, while their neighbors at Natchitoches enjoyed a wider selection of goods, obtained cheaply from New Orleans via the water route up the Mississippi to the Red River. Not surprisingly, such edicts did not curtail illicit trade between Nacogdoches and Natchitoches. Livestock flowed across the ill-defined boundary from Texas into Louisiana and much-needed food and consumer goods found their way back from Natchitoches.

Immediately after learning of the American acquisition of Louisiana, the Spanish government again ordered a halt to all trade between Nacogdoches and Natchitoches. The sale of always abundant livestock across the boundary to Americans even became an offense punishable by death. Yet local officials at Nacogdoches generally refused to enforce the ban, because the people of the District of Nacogdoches had always relied on Natchitoches for their survival and economic livelihood. Despite local leaders'

willingness to disobey crown law and trade across the international boundary, once the United States took possession of Louisiana, the economic, social, and cultural networks and interdependence of the Texas-Louisiana borderlands started breaking down. After 1803 tensions involving runaway slaves, Indian trade, and the unsettled location of the international border drove Spain and the United States to the brink of war. The major disagreement focused on the border between the two countries and on control of the lands between the Arroyo Hondo—a stream flowing between Natchitoches and the site of the abandoned Spanish presidio of Los Adaes—and the Sabine River.

Prior to the Purchase, the United States had posed no threat to Texas, even though Spanish officials believed otherwise. During the 1790s Irish adventurer Philip Nolan, a former associate of nefarious American General James Wilkinson, had secured passports from successive Spanish Louisiana Governors Esteban Miró and the Baron de Carondelet to conduct trade in East Texas. He traveled to Nacogdoches, San Antonio de Bexar, and to Nuevo Santander rounding up horses that he drove to Louisiana. Yet some Spanish officials, such as the governor of Natchez, believed Nolan and other American agents were inciting Texas Indians against Spanish rule. Leaving Natchez again in October 1800, this time without a passport to enter Texas, Nolan and his followers established a small fort and horse corrals north of Nacogdoches. Spanish Commandant General Pedro de Nava, convinced that Nolan was an American agent, dispatched troops from Nacogdoches in March 1801. The Spanish forces killed Nolan, captured his associates, and thereby brought an end to the threat.

The first real post-Purchase problem along the Texas-Louisiana border concerned a number of Louisianans of French and Spanish descent who wanted to immigrate to Texas rather than become United States citizens. Spanish officials expressed differing opinions on whether these foreigners should be permitted to enter Texas. Governor Juan Bautista de Elguezabal wanted to admit them, arguing they could serve as a buffer against the expansionist Americans and also help defend the province. Commander-General Nemensio Salcedo disagreed, believing that these immigrants would not swear loyalty to Spain; even if permitted to enter Texas, he argued, they should be settled well into the interior. Salcedo's distrust prevented the governor from settling a thousand Louisiana families near the mouth of the Trinity River.

The Texas-Louisiana border dispute represented an inevitable conflict between Spain and the United States that surprisingly did not happen sooner. The three years following the American occupation of Louisiana witnessed bellicose posturing between border posts but no engagements. In one instance an American expedition repulsed a Spanish force without bloodshed, prompting Spanish troops to withdraw from posts in the disputed territory claimed by the United States. By 1805 the Spanish had increased the troop presence in East Texas with 141 stationed at Nacogdoches to deal with the American menace. Yet this resulted in more Spanish patrols crossing the Sabine. Officials at Natchitoches complained to the Spanish commander at Nacogdoches that troops illegally crossed the Sabine boundary into American territory. The Spanish denied these allegations, claiming instead that the patrols protected territory belonging to Spain.

The Spanish moves prompted American reactions. In early February 1806, United States troops and armed volunteers approached the Spanish outpost east of the Sabine at the former presidio of Los Adaes and demanded that the Spaniards withdraw. Outnumbered, the Spanish patrol left without a fight. War appeared inevitable. Lieutenant Colonel Simón de Herrera, the former Governor of Nuevo León, assumed command of Spanish forces in East Texas in June 1806, as war rumors circulated on both sides of the Sabine. Spanish reports credited American General James Wilkinson with as many as fifteen thousand men at Natchitoches. Louisiana Governor William C.C. Claiborne insisted that the Spanish had twenty thousand troops along the Sabine. In reality, the Spanish had some eight hundred soldiers, representing the largest detachment ever stationed in East Texas. Conversely, Wilkinson's army at Natchitoches had no more regular troops than the Spanish.

Neither Spain nor the United States wanted war. The two nations ultimately avoided hostilities when Wilkinson informed the Spanish on October 29, 1806, that he would withdraw American troops east of the Arroyo Hondo, if the Spanish withdrew west of the Sabine. Herrera agreed, and the territory between the rivers became a neutral ground with neither side permitting settlers nor troops there until diplomats settled the boundary. This agreement, often referred to as the Neutral Ground Treaty, is incorrectly designated since neither government ever ratified it. Still, both sides did abide by its provisions until 1821, and the existence of this buffer zone prevented any serious international conflict. Even so, as a land outside United States or Spanish control, the Neutral Ground soon became a lawless haven for adventurers, filibusters, and an assortment of other undesirables causing trouble on both sides of the border.

PLATE 11

Première Partie de la Carte de l'Intérieur de la Louisiane by Zebulon M. Pike. Published in *Voyage au Nouveau Mexique*: Paris, 1812. Drafting by Antoine [Anthony] Nau. 18" x 18' 2".

This map is from the French edition of Pike's 1810 An Account of Expeditions to the Sources of the Mississippi and through Western Parts of Louisiana. . . During the years 1805, 1806, and 1807 etc. After the 1803 Louisiana Purchase, President Thomas Jefferson dispatched Lewis and Clark to the Northwest Territory and Zebulon Pike southwest to explore the acquired possessions. Pike's survey traced the Arkansas and Red rivers to their sources. Pike's Texas cartography was firsthand and more accurate than Alexander Von Humboldt's 1810 map of New Spain. Humboldt later accused Pike of plagiarism. Pike's published reports predated Lewis and Clark's by four years. This is the earliest reliable map of east and northern Texas.

The expedition of Captain Zebulon Pike in 1806-1807 provided the only other serious dispute between Spain and the United States concerning Texas during the next few years. The Pike expedition, supposedly an exploration of present-day Colorado, alarmed Spanish officials who believed he sought lands favorable for American expansion. Acting on their suspicions, the Spanish arrested Pike's party and escorted them to Santa Fe and later to Chihuahua City where they met with the governor. After a brief delay, the Spanish escorted Pike's expedition back through Texas to Nacogdoches and released them (Plate 11).

The Mexican Revolution proved the final crisis for Spanish Texas, when on September 16, 1810, Father Miguel Hidalgo y Costilla issued a call for independence. Although his plea occurred far from Texas, Juan Bautista de las Casas and other conspirators used the appeal to seize temporary control over San Antonio. The seige was short-lived and by the summer of 1811 Spanish loyalists had regained power in Texas. A more serious challenge came with the José Bernardo Maximiliano Gutiérrez de Lara-Augustus William Magee filibuster of 1812-1813. The filibuster army—which included Americans, Frenchmen, and Texas revolutionaries—organized at Natchitoches with American assistance, crossed the Sabine on the well-worn Camino Real and advanced on Nacogdoches, where they found a warm welcome. Although the Spanish commander tried to rally resistance, the ten soldiers of the Nacogdoches garrison and a number of citizens instead joined the expedition.

From Nacogdoches the filibusters moved south, capturing La Bahia and San Antonio de Bexar, before proclaiming Texas an independent state in the spring of 1813. Gutiérrez quickly

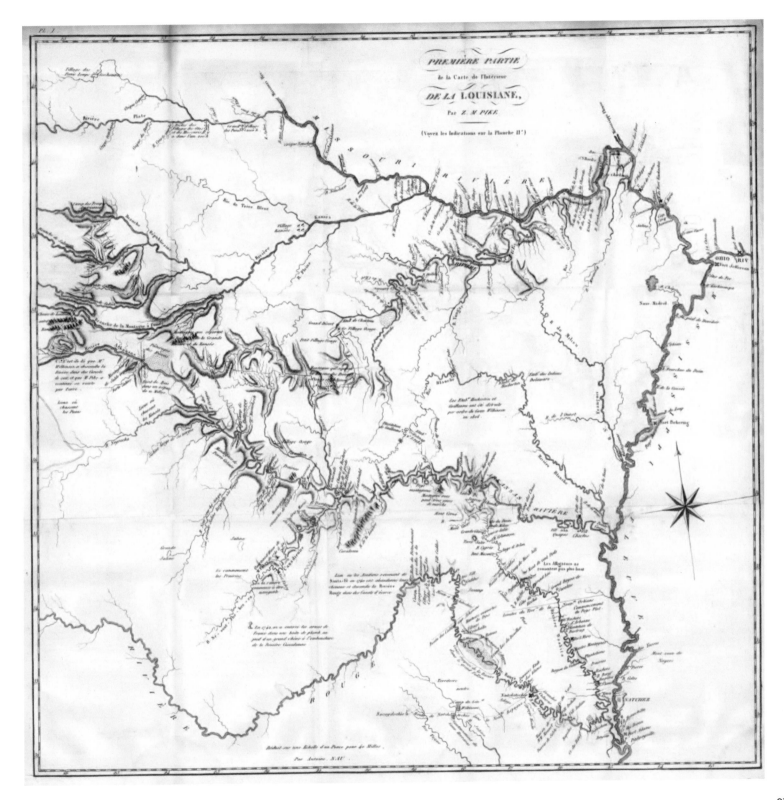

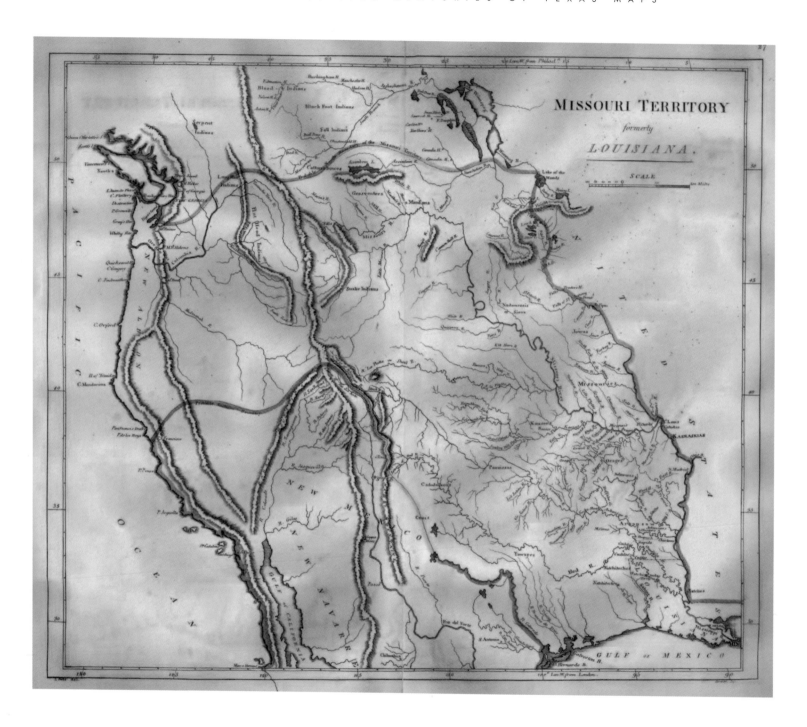

PLATE 12

Missouri Territory (formerly Louisiana) by Mathew Carey: Philadelphia. 1814. Lithograph by Bower; drafting by Samuel Lewis. 12" x 14".

The map incorporates information from the Lewis and Clark Expedition. Samuel Lewis, the map's creator, used William Clark's manuscript map, not then ready for publication. The Missouri Territory extends from the Mississippi River to the Pacific, with probable boundaries indicated by dotted lines. The province of Texas is shown below the southern boundary of the Colorado River, "Arkansaw River" and "Stoney Mountains."

drafted a constitution. He also executed Spanish prisoners, which alienated Americans and prompted them to quit the revolutionary cause. Then in August, a royalist force led by José Joaquín Arredondo routed the reduced filibuster army at the Battle of Medina. Arredondo immediately restored royal authority through confiscation, detention, and execution, shooting some 327 people in San Antonio. Panic-stricken, many of the people of Nacogdoches fled across the Sabine River to friends, relatives, and safety in American territory in time to avoid a bloody purge by one of Arredondo's lieutenants. The severity of the royalist retaliation momentarily ensured that Texas remained in Spanish hands.

After 1813 Spanish officials chose to focus on Texas rather than Florida, as the region's geopolitical importance had greatly surpassed its economic or demographic significance—Texas had replaced Florida as the defensive bulwark of Spanish North America. Texas sat adjacent to Spain's rich Mexican colony while the isolated Florida peninsula, by contrast, was slipping piece by piece from Spanish control. Additionally, in September 1810 the Baton Rouge Revolution divested Spain of the West Florida parishes of Louisiana. In March 1812 an American-led "patriot" expedition attacked Spanish East Florida settlements, capturing all but St. Augustine before the United States withdrew support to avoid fighting a war with both Britain and Spain. In April 1813—during the War of 1812 against Britain—General Wilkinson captured Spanish Mobile, claiming that the area between the Pearl and Perdido rivers had been part of the Mississippi Territory. The following year and again in 1818, United States General Andrew Jackson seized Spanish Pensacola, demonstrating how little Spanish authority existed in the isolated province of Florida. During the summer of 1817 Scotsman Gregor MacGregor and a group of filibusters established the short-lived Republic of Florida on Amelia Island, further illustrating Spain's impotence in the peninsula.

Although not sponsored by the United States, filibuster and revolutionary activity against Texas also increased after 1815,

even though these efforts lacked a single leader or single objective. Some expeditions wanted Mexican independence while others preferred American or even French occupation. Frenchman Luis Aury, who had temporarily occupied Amelia Island, arrived at Galveston in the fall of 1816, establishing a base for the Mexican independence movement rather than for an American effort to seize Texas. But he soon departed, leaving Galveston under the control of Jean Lafitte. During the spring of 1817 former Spanish patriot and liberal Francisco Xavier Mina arrived at Galveston with troops for the invasion and liberation of all Mexico, but they departed less than a month later. A group of French exiles, under the leadership of brothers Charles François Antoine and Henri Lallemand, launched one of the more ambitious plans involving Spanish Texas. Negotiating with the Spanish to provide a military force against American expansion while also treating with the United States to provide a forward base for American designs on Texas, they established the short-lived settlement of Le Champ d'Asile on the Trinity River. Poorly supplied and lacking additional volunteers, the Frenchmen retreated to Galveston in July 1818, ending French attempts to settle in Texas. The last filibuster against Spanish control began during the summer of 1819 when American James Long attempted to capture Texas. Well equipped and provisioned, the expedition lasted in an irregular off-and-on fashion from the summer of 1819 until the fall of 1821, when Long finally surrendered. His capitulation and supposed accidental death in Mexico City marked the end of early filibustering into Texas.

In February 1819, after three years of negotiation, the United States and Spain signed the Adams-Onís or Transcontinental Treaty, in which Spain ceded all of the Floridas to the United States. In return Spain retained Texas. The United States agreed to the Sabine and Red rivers as the southern boundary of the Louisiana Purchase, and the Neutral Ground officially became part of Louisiana. The United States temporarily relinquished its claim to Texas, but in doing so reinforced its claim to the Pacific. But just as Arsène Lacarrière Latour had predicted in 1816, the time had come, "when the Americans . . . will pour in myriads into Mexico." That onslaught began in earnest during the 1820s as an independent Mexico attempted to accomplish what Spain could not. Establishing an empresario program, Mexico opened the borders and Americans streamed across the Sabine River into Texas. By the mid 1830s Mexico faced the full brunt of American Manifest Destiny. During the next twenty years, the nexus of empire vanished as the U.S seized lands from Texas across to California (Plate 12).

SUGGESTED READINGS

Donald E. Chipman. *Spanish Texas, 1519-1821*. Austin: University of Texas Press, 1992.

James E. Lewis. *The American Union and the Problem of Neighborhood: The United States and the Collapse of the Spanish Empire, 1783-1829*. Chapel Hill: University of North Carolina Press, 1998.

Frank L. Owsley, Jr., and Gene A. Smith. *Filibusters and Expansionists: Jeffersonian Manifest Destiny, 1800-1821*. Tuscaloosa: University of Alabama Press, 1997; reprint 2004.

Harris Gaylord Warren. *The Sword was Their Passport: A History of American Filibustering in the Mexican Revolution*. Baton Rouge: Louisiana State University Press, 1943.

William Earl Weeks. *John Quincy Adams and American Global Empire*. Lexington: University Press of Kentucky, 1992.

Bruce Winders

COLONIZATION
1821-1836

RICHARD BRUCE WINDERS

Texas changed nationality three times between 1821 and 1836. Initially Spanish, Texas was briefly under Mexican rule in 1821. By 1836, Texas had ventured on the new path as an independent republic. This brief fifteen-year-period set in motion events which made possible Texas' admission to the United States in 1845.

Spain's grasp on New Spain, of which Mexico and Texas were a part, had been slipping for some time. Napoleon Bonaparte's occupation of Spain and the imprisonment of King Ferdinand VII in 1808 loosened that country's hold over its colonial holdings in the Western Hemisphere. The king's supporters throughout Latin America formed *juntas* and, by ruling in his name, attempted to keep the connection intact. Ferdinand regained his throne in 1814 and resumed his absolutist style of governing his kingdom.

The political climate in the distant territory had changed during Ferdinand's absence. The temporary break with the Spanish homeland had given its colonies a taste of independence—and they liked it. Many Spanish colonies began to believe that the king needed them more than they needed him, and the existence of a fledging republic called the United States of America to the north demonstrated that political independence was possible.

Something else had changed as well: resentment replaced dutiful respect for the Spanish. Society had developed into a distinct caste system under the Spanish. Special privileges and the highest positions were reserved for *Peninsulares* or Spanish-born subjects living in the colonies. Below them were the *Criollos*, the colonial-born offspring of Spanish parents. Although of Spanish ancestry, *Criollos* were considered socially inferior to *Peninsulares*. Lower down still were the mixed bloods (*Meztisos*), Indians, and Africans.

In 1810, anger against the Spanish in Mexico exploded with Father Miguel Hidalgo y Costilla's *"Grito de Dolores."* Although originally a *Criollo* plot to gain equality with *Peninsulares*, his revolt unleashed lower-class rage against all things Spanish. Hidalgo's followers made no distinction between *Criollos* and *Peninsulares*, causing these two factions to band together for survival. In 1821, however, Colonel Agustín de Iturbide, a *Criollo* serving in the royalist cause, formulated a plan that enabled all Mexicans to unite briefly in order to achieve their independence from Spain. He rallied his countrymen around three principles: independence from Spain, protection of the Catholic Church, and an equal union of all Mexicans. Iturbide's *Plan de Iguala* allowed Mexicans to end Spain's three-hundred-year reign over their lives.

These events had ramifications for Texas. In 1811, Capt. Juan Bautista de las Casas convinced the militia of San Antonio de Béxar to rebel in support of Hidalgo's revolt. Local officials quickly put down Las Casas' revolt, but the following year brought an even larger disturbance to Texas. In 1812, José Bernardo Gutiérrez de Lara, a supporter of Hidalgo, and Augustus W. Magee, a former United States Army officer, joined together to lead a filibustering expedition from Louisiana into Texas. The expedition ultimately captured San Antonio de Béxar in April 1813 and announced the establishment of the Green Flag Republic. The republican endeavor was smashed by a royalist counterattack, which culminated in a decisive royalist victory at the Battle of Medina on August 18, 1813. Other intrigues followed on a smaller scale, but Texas stayed relatively quiet after Spanish officials reasserted their authority.

Mexico's first attempt to create a government following independence was hijacked by Agustín de Iturbide. The *Plan de Iguala* called for the establishment of a constitutional monarchy

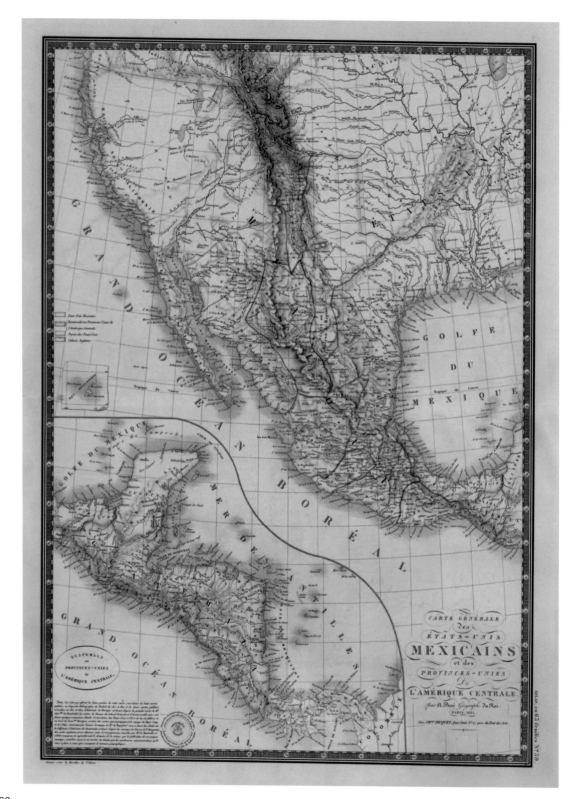

PLATE 14

Carte Générale des États-Unis Mexicains et des Provinces-Unies de l'Amérique Centrale by Adrien H. Brué. Published by Charles Picquet in *Atlas Universel de Géographie, Physique, Politique et Historique, Ancienne et Moderne*; Paris, 1825.

Charles Picquet acquired all of Brué's plates after his death and reissued this edition in 1835. Jedidiah Smith's western discoveries of the 1830s are not incorporated in this edition. The map follows Mexican independence from Spain on August 24, 1821. Northwest Mexico extending to California and Oregon is described as "Unknown Country." Texas has become a separate province of Mexico.

PLATE 16

A Map of the United States of Mexico as Organized and Defined by . . . Congress of that Republic by Henry Schenck Tanner. Published by H.S. Tanner in Pocket Map in Boards, "Mexico"; Philadelphia, 1834. 23" x 29".

This large pocket map includes two insets ([1] Table of Distances [2] Map of Roads from Vera Cruz to Mexico). With an April 2, 1832, copyright but an 1834 edition, this is commonly considered the source for John Disturnell's plagiarized "Treaty Map," used in 1846. Tanner's original sources were Humboldt, Darby, and Pike. Five editions of Tanner's popular map were issued between 1825 and 1847. Early editions showed the boundary between Lower and Upper California 120 miles south of San Diego, an error corrected by the Treaty of Guadalupe Hidalgo and the Gadsden Purchase. This edition used information from the more famous Austin-Tanner map of Texas.

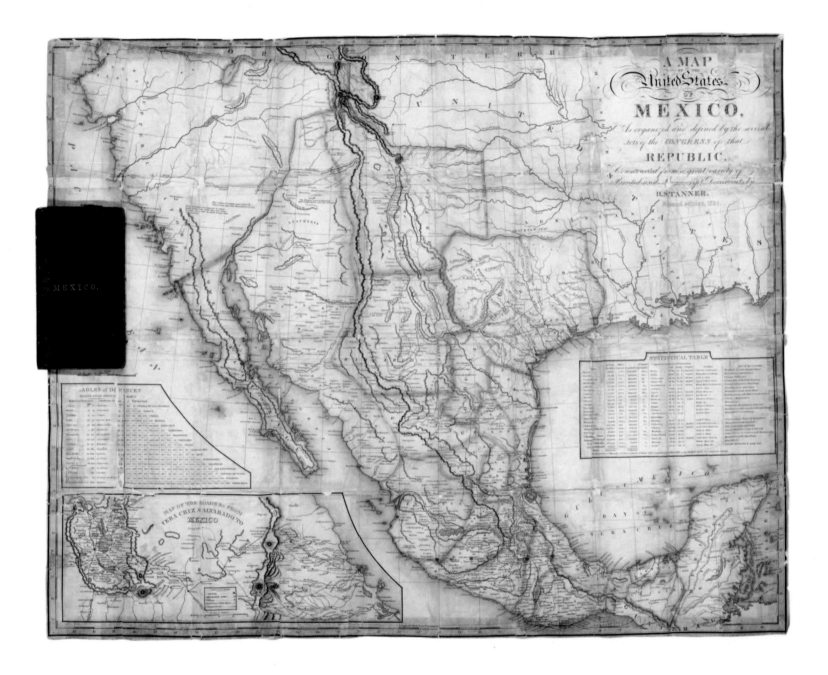

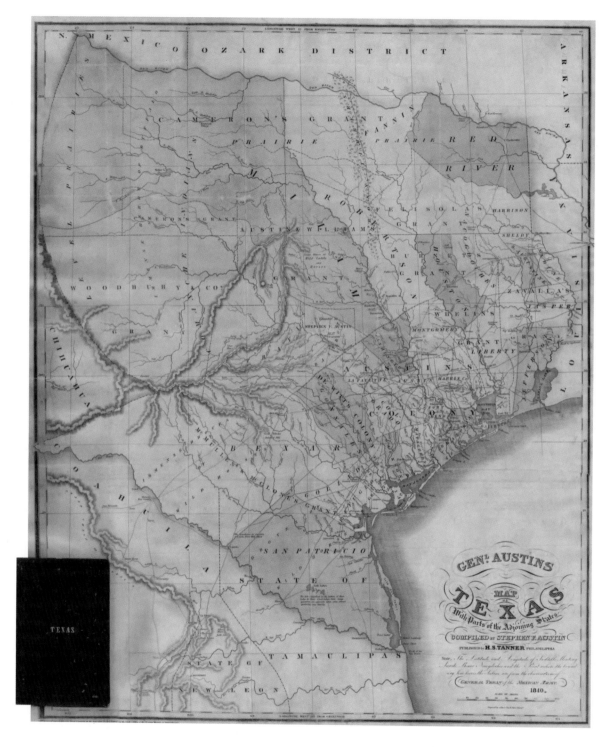

PLATE 21

Genl. Austin's Map of Texas with Parts of Adjoining States . . . From the Observations of General Téran of the Mexican Army by Henry S. Tanner. Published in Pocket Map *Texas*: Philadelphia. 1840. Compiled by Stephen F. Austin with engineers John and William W. Warr. Philadelphia. 28.5" x 23.3".

The first Austin-Tanner map was printed in 1830. Seven editions followed. This statehood version also was used in Francis Moore's "Map and Description of Texas" (Philadelphia, 1850). The City of Austin is added in this edition. Ozark territory is shown to the north. Streeter considered this the cornerstone piece of any Texas collection and perhaps the most important map in Texas history. The provenance of this copy can be traced through Randolph Bryan, son of Moses Austin Bryan.

with a European nobleman at its head. Iturbide, supported by his troops, forced the nascent government to offer the crown to him, which he accepted on May 18, 1822. Forced from power in less than a year, Iturbide nevertheless had unintentionally opened the door to the formation of the Federal Republic of Mexico (Plate 14).

Latin Americans, including Mexicans, had been exposed to the same ideas of the Enlightenment as had the inhabitants of the English colonies to the north. A republic offered an alternative to the monarchy and reflected the emerging notion of the intrinsic rights of man. Mexican republicans used Iturbide's unsuccessful reign to propose that the time had come for Mexico to join the increasing ranks of republican nations. In 1823, delegates met to transform Mexico into a republic, an act that proved to be much more difficult than they imagined. They sealed this political change with the adoption and implementation of the Federal Constitution of 1824.

The new constitution had a pronounced effect on Texas. Texas under the Spanish had been governed as a separate province within New Spain. Directed by the Federal Constitution of 1824, however, Texas was attached to the former province of Coahuila to form a new state called Coahuila y Tejas. Coahuila received the greater leadership role of the two since the capital was to be located in Saltillo instead of San Antonio de Béxar. The reason for the new arrangement was that Texas was deemed to lack a sufficient population for statehood. Residents of Texas, although supporting the new government, saw the act as a demotion and desired separate statehood within the federation (Plate 16).

To be fair, Texas did have a relatively small population for a region so big, even if one discounted the numerous bands of Indians. The Spanish had successfully used missions and presidios to establish viable beachheads on the frontier, and the fruits of these labors had produced three population centers: San Antonio de Béxar, Nacogdoches, and Goliad. Still, the Spanish wished to

PLATE 17

Map of the State Coahuila and Texas by William Hooker. Published by Goodrich & Wiley in *A Visit to Texas . . . .*: New York, 1834.

First issued in Mary Austin Holley's book Texas. Observations, Historical. . . In Autumn of 1831, *Armstrong & Plaskitt, (1833), this map is a smaller reprint of her cousin Stephen Austin's map because Tanner refused permission to use his copyrighted edition. The map clearly identifies empresario grants just prior to revolution while Texas was still a province of Mexico. The border is shown east of the Rio Grande River. The northern border is limited by "Oregon Territory." Western Texas cartography is scant.*

PLATE 18

A New Map of Texas with the Contiguous American & Mexican States
by James Hamilton Young. Pocket map published by S. Augustus
Mitchell in "Mitchell's Map of Texas"; Philadelphia, 1835. 13" x 15".
*One of the most influential of the Republic of Texas maps, just predating
the revolution by months, this was produced ultimately in eight editions
through statehood. The map corrected earlier placements of Texas-
Louisiana borders by moving half a degree to the east. It shows Mexican
land grants to encourage American settlement. The western border
extends only to Nueces River. Land north of the Red River is described
as "Santa Fe formerly New Mexico," resulting in so-called "strawberry
shaped" Texas. In the remark section Texas is described as "finest stock
countries in the world." Politically it goes on to predict that when the
population reaches "50,000, the people will endeavor to obtain a government separate from that of Cohahuila (sic) and to establish their own
legislature at San Felipe with representatives to the Mexican Congress."*

boost the number of civil settlements in Texas. The answer to
the problem seemed to be delivered to them by Moses Austin
who requested permission to bring American colonists to Texas.
Mexico inherited the sparsely inhabited frontier along with the
problem of how to populate it. Thus, the new Mexican nation
turned to colonization as the means to bring growth and prosperity to Texas.

Colonization was to be conducted in an organized
manner. Laws were passed, first by the empire and then the
republic, which laid out the framework for what essentially was
an immigration policy. Lawmakers decided to conduct colonization through land contractors or empresarios. Each contractor
received a designated territory or colony into which he promised
to settle a specified number of families or individuals. Colonists
were to be screened in order to prevent men and women of bad
character from entering Texas. Those persons accepted were
required to adopt Mexican citizenship as well as the Catholic
religion. In return, colonists would receive land through their
respective empresario for their use.

Although at first orderly, colonization quickly grew out
of control. In 1823, only several hundred Americans were living
in Texas; by 1828, the number was approaching 30,000. Some
early empresarios, like Stephen F. Austin, took their obligations
seriously and acted in the best interest of their colony and the
national government. Others used their grants to promote land
speculation in the United States, forming emigration companies
intended to sell Mexican land to United States citizens and
facilitate their removal to Texas. Thus by 1828, the American

population living in Texas was a combination of legal and illegal
immigrants, which did not sit well with the Mexican government.

Several factors contributed to the rapid influx of Americans to Texas. In 1820 the United States Congress revised the
nation's public land laws. The new regulations called for public
land to be auctioned in lots of eighty acres each. The cost per
acre was set at $1.25, bringing the purchase price of eighty acres
to $100. In addition, lawmakers eliminated purchases on credit
and stated that buyers must pay in specie (gold or silver). These
changes were intended to streamline the dissemination of public
land, but they benefited speculators more than farmers. Under
Mexico's colonization laws, immigrants received use of the land
while only having to pay an empresario twelve-and-a-half cents
per acre to offset the land agent's expense of survey and record-
ing the grant. Americans also took comfort in the idea that they
would be living in a federal republic, a political system they
strongly supported.

Empresarios recruited colonists a variety of ways. Many
sent agents to the United States or ran advertisements in United
States newspapers. Stephen F. Austin believed that prospective
colonists needed to see where they would be living. His detailed
manuscript map of Texas was printed and distributed by Philadelphia map publisher Henry S. Tanner (Plate 21). Empresarios
Joseph Vehlin, David G. Burnet, and Lorenzo de Zavala formed
the Galveston Bay and Texas Land Company, which published a
handbook entitled *Guide to Texas Emigrants*. By land and sea,
Americans flooded into Texas in an uncontrollable torrent.

Recruitment of Americans was cut short in 1830 when
the Mexican government attempted to reassert its authority over
immigration to Texas. The government had been driven to action
after reading General Manuel de Mier y Terán's report of his
1828-1829 visit to Texas. The general announced that Texas was
on the verge of being lost to Mexico through the unexpected number of Americans who had recently arrived. Instead of assimilat-
ing into Mexican society and adopting the native culture, many
Americans continued to live as they had in the United States.
A review of their newly established communities revealed names
like Washington, Columbia, Liberty, and Harrisburg. He was
certain that Americans would wrest Texas from Mexico unless
steps were taken immediately.

Mexico's response to Mier y Terán's warning was the Law
of April 6, 1830. The goal was to regain a firm grip on Texas by
bringing immigration under control. Colonists from the United
States could no longer immigrate to Texas. Existing garrisons
were to be reinforced as well as new ones established. Tariffs,

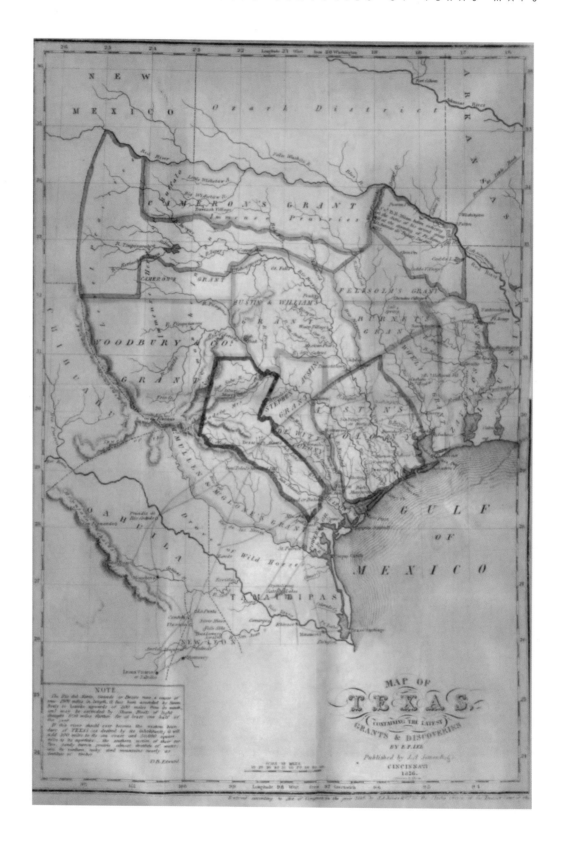

PLATE 19

Map of Texas Containing the Latest Grants & Discoveries by E. F.
Lee. Published by J. A. James & Co.in *The History of Texas . . .* by
David B. Edwards: Lithography by Doolittle & Munson: Cincinnati.
1836. 9" x 13".

One of the last Texas Mexican provincial maps, this is a folding, litho-
graphed, hand-colored map with Mexican land grants. The text in the
accompanying book details colonization regulations. The author lived
in Gonzales, Texas, from 1830 through the revolution.

from which colonists had been exempted, were to be collected
by revenue agents who would soon arrive. Slavery was prohibited
in the belief that the lack of labor would make Texas unattractive
to Americans from the slave states. Unfortunately for Mexico, the
Law of April 6 failed to halt events that were already in progress,
both inside Texas and in the Mexican nation as a whole.

Independence had not brought stability to Mexico. The
political alliance between various factions, brought together by
the *Plan de Iguala*, quickly shattered. Two parties emerged that
would vie for control of the government for decades to come.
The federalists supported the federal republic and the Constitution
of 1824, which had created a role for the states. The centralists
disliked the dispersal of power, desiring instead a centralized
bureaucracy similar to what Mexico had during Spanish rule.
While Texas was being colonized, Mexico experienced a series of
revolts and counter revolts between the federalist and centralists
which brought to power a strongman from the state of Vera
Cruz—General Antonio López de Santa Anna.

Santa Anna had pushed his way to the top to become
Mexico's premier political chieftain. As a youth he had disdained
commerce and entered the Spanish army, first serving near Vera
Cruz, the region of his birth. The young lieutenant accompanied
the royalist force sent to defend Texas from the threat posed by
Gutiérrez and Magee, earning a citation for bravery at the Battle
of Medina in 1813. The *Criollo* officer returned to Vera Cruz
where he steadily advanced in rank over the next few years. He
grew tired of garrison duty, though. Iturbide's pronouncement for
independence appealed to the ambitious Santa Anna, to whom
Iturbide offered a general's star in return for his support. A dispute with his benefactor caused him to work for the emperor's
ouster and the establishment of the republic. In 1829, Santa
Anna acquired the sobriquet of "Hero of Tampico" for stopping a
Spanish attempt to reclaim Mexico. In 1832, Santa Anna earned
the admiration of Mexican federalists throughout the country
when he led a counter revolt against a centralist administration

then in power. This wave of popularity swept Santa Anna—savior
of the republic from both external and internal enemies—into the
office of the president.

Even Texans supported Santa Anna in his rise to the presidency. In June 1832, a young Texas lawyer named William B.
Travis instigated a revolt against the centralist garrison at
Anahuac, a revolt his supporters claimed was carried out to aid
Santa Anna's activities deeper in Mexico. Many of the men emerging as leaders in the American colonies believed that Santa Anna,
whose actions so far had shown him to be an ardent federalist,
would grant Texas separate statehood once he was in power. They
and others would be disappointed by what lay ahead (Plate 17).

Santa Anna soon abandoned the federalist cause once he
took office. Republican reform threatened the special position
held by the army, the church, and landowners, who warned they
would withdraw their support unless he would protect their interests. By 1835, Santa Anna had switched allegiance, disbanded
congress, and announced that the Constitution of 1824 would
no longer be followed. In essence, all states within the Mexican
Federation were being stripped of statehood and converted into
districts that would be managed from Mexico City. Mexican
republicans, shocked and angered by Santa Anna's centralist
conversion, prepared to resist the man who had betrayed them.

Opposition to the centralists tended to be strongest in the
states some distances from central Mexico. The reason was in the
nature of federalism: statehood allowed for a measure of local
control not found in centralism. This was especially important in
a time when transportation and communication were extremely
slow and difficult. Frontier regions like Texas faced different
problems than those clustered in Mexico's heartland. The only
hope for Texas' statehood now was to defeat the centralists and
restore the Federal Constitution of 1824.

Revolution was not an unknown concept to Texans, since
Americans living in Texas had grown up on tales of their fathers'
fight against the British and many Tejanos had either partici-
pated in or witnessed the Las Casas and Green Flag revolts.
Tejanos and the colonists had been drawn together in an alliance
to press for Texas' statehood. The question was how to proceed:
cooperate with other Mexican republicans in the fight against
Santa Anna or go it alone (Plate 18).

Independence had not been the Texans' initial reason for
rebellion. In November 1835, Texas' revolutionary government
had issued the Declaration of Causes. The document announced
that the revolt's objective was to reestablish the Federal Constitution of 1824 with separate statehood for Texas within the

restored Mexican Federation. However, a number of rebels were already prepared to sever all ties with Mexico, believing they had the right to separate from a nation with whom they shared few interests. During the early months of the revolution an attempt was made to build a united front against the centralists by coordinating with opponents in other Mexican states. Once the hoped-for coalition with Mexicans federalists failed to materialize, however, the independence party gained control of the revolution's direction and rejected any further connection with Mexico.

The Texas Revolution lasted only about six months from start to finish. In early October 1835, the colonists resisted an attempt by troops loyal to the government to reclaim a small cannon given to the town of Gonzales. Shots were fired, announcing Texas' intention to support the general uprising. Successes came quickly to the rebels. By December they had captured Goliad and Béxar and driven all centralist garrisons from Texas soil. The tide turned in February 1836 when Santa Anna mounted a two-pronged counterattack into Texas aimed at retaking Goliad and Béxar. After crushing the rebels at these locations, Santa Anna turned eastward in an effort to drive the Americans and anyone who supported them out of Texas. On April 21, Santa Anna's plans came to a sudden halt when General Sam Houston brought him to bay along the banks of the San Jacinto River and handed him a crushing defeat. Although 4,000 Mexican troops still remained in Texas, Santa Anna's defeat at San Jacinto effectively secured Texas' independence, which had been declared on March 2, 1836.

In October 1836, Texans—now independent of Mexico and citizens of the new Republic of Texas—elected General Sam Houston as president of the new republic. The voters also expressed an overwhelming desire to join the United States,

something that many assumed would immediately follow. Texas was destined, nonetheless, to remain independent for nearly ten years before becoming the twenty-eighth state in the Union on December 29, 1845.

The Texas Revolution changed the map of Texas (Plate 19). During Spanish rule, Texas had been much smaller in size, with its southern boundary terminating near the Nueces River. Following Houston's victory at San Jacinto, Santa Anna had agreed to order all Mexican troops still in the field to withdraw to the Rio Grande. The new government of the Republic of Texas afterwards claimed the Rio Grande as the southern border instead of the Nueces. The border gave Texas claim to much of New Mexico since the river runs west of the town of Santa Fe. The United States recognized this expanded version of Texas when it was annexed. Texas' boundaries would not change until the Compromise of 1850 when the state assumed its modern shape.

SUGGESTED READINGS

Cantrell, Gregg. *Stephen F. Austin: Empresario of Texas.* New Haven: Yale University Press; New Edition, 2001.

Holley, Mary Austin. *Texas.* Austin: Texas State Historical Association, 1985.

Owsley, Frank L. and Gene A. Smith. *Filibusters and Expansionists: Jeffersonian Manifest Destiny, 1800-1821.* Tuscaloosa: University Alabama Press; New Edition, 2004.

Tijerina, Andres. *Tejanos and Texas Under the Mexican Flag 1821-1836.* College Station: Texas A & M Press, 1994.

Winders, Richard Bruce. *Crises in the Southwest: The United States, Mexico, and the Struggle over Texas.* Wilmington: Scholarly Resources, 2002.

THE REPUBLIC OF TEXAS
1836 TO 1845

GREGG CANTRELL

The Republic of Texas came into being on March 2, 1836, when fifty-nine men affixed their signatures to the Texas Declaration of Independence in the frontier village of Washington-on-the-Brazos. It would be hard, however, to imagine less auspicious circumstances for the birth of a new nation. The fledgling republic was penniless, it had no functioning government, and, less than one hundred miles away, a large Mexican army was laying siege to the Alamo. Few objective observers would have given the new republic very good odds of securing its independence.

Yet the Republic of Texas did survive. Over the next two weeks, the delegates at Washington-on-the-Brazos forged a new government and named Sam Houston, a former governor of Tennessee, commander of all Texas military forces. When news arrived of the fall of the Alamo and the slaughter of captured Texan troops at Goliad, Houston managed to gather most of the remaining able-bodied men into a ragtag army and began a tactical retreat into East Texas. On April 21, he defeated President-General Antonio López de Santa Anna's troops at San Jacinto, and independence was won.

But the improbable victory at San Jacinto by no means secured the future of the new republic. The Texans still faced a host of daunting challenges. The military triumph of April 21 came only because Santa Anna imprudently had personally taken a fraction of his troops deep into the heart of the Anglo settlements in pursuit of Houston's army, leaving behind the Mexicans' sources of supplies and two-thirds of the Mexican army. Fortunately for the Texans, Santa Anna was taken prisoner on April 22, and the captive president ordered his remaining troops to withdraw from Texas. Demoralized by their commander's defeat and weakened by torrential rains that turned southeast Texas

into a "sea of mud" (to use one Mexican general's description of conditions), Santa Anna's generals complied with the orders to retire south of the Rio Grande.

Even with Texas cleared of Mexican troops, the future in the spring of 1836 did not look promising. Sam Houston had been wounded at San Jacinto and was forced to travel to New Orleans to seek medical attention. The interim government headed by President David G. Burnet was weak and discredited, and the unruly Texan army could not be controlled by anyone but Houston. The threat of a re-invasion by Mexico was by no means gone. Hostile Indians and a Tejano population with uncertain loyalties compounded the security problem. Diplomatic recognition by the United States and the major powers of Western Europe would not be forthcoming anytime soon. Towns and farms had been laid waste by the destructive campaigns of the past year, and much of the Anglo population had fled to Louisiana to escape Santa Anna's onslaught. Many of the men who had fought for the Texan cause were newcomers to Texas who had come in pursuit of military glory or quick riches. Indeed, of the fifty-nine signers of the Declaration of Independence, only ten had been in Texas more than three months, and only two—José Antonio Navarro and José Francisco Ruiz—were native Texans (Plate 22).

Even the boundaries of the Republic were uncertain. Maps from the Spanish and Mexican eras generally show the Nueces River as the border between Texas and the provinces to the south, with the ill-defined western boundary between Texas and New Mexico falling well to the east of modern state lines. But in the Treaties of Velasco, agreed to by Santa Anna shortly after his capture, the Texans claimed the Rio Grande, from the Gulf of Mexico to its source in southern Colorado, as the southern and western boundary of Texas. And if that were not enough, the

PLATE 22

A New Map of Texas–1841
by Day & Haghe. Published
by Smith Elder and Company,
Cornhill, in *A History of the
Republic of Texas* by Nicholas
Doran Maillard; London, 1841.
16.7" x 15.2"

*The legend color outlines the
political, conventional, and national "Boundaries of Texas from
1819 through the revolution."
Also noted are Indian tribes,
cities, and rivers. The author,
an attorney, spent six months
in Texas in 1839, before leaving
as no friend of the new Republic.
He characterized the infant
nation as "a country filled with
habitual liars, drunkards,
blasphemers, and slanderers,
sanguinary gamesters and
cold-blooded assassins." He de-
nounced the Republic in general
and its inhabitants in particular.
He referred to Stephen F. Austin
as "the prince of hypocrites" and
James Bowie as a "master."
Maillard describes himself as
"an impartial historian."*

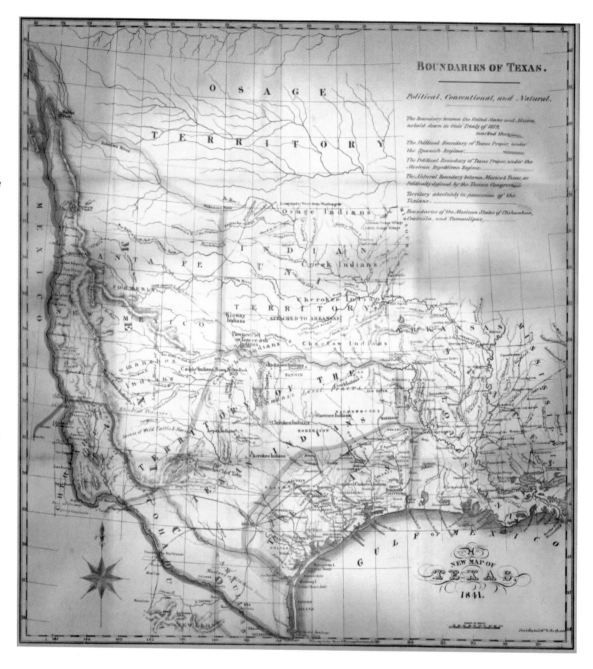

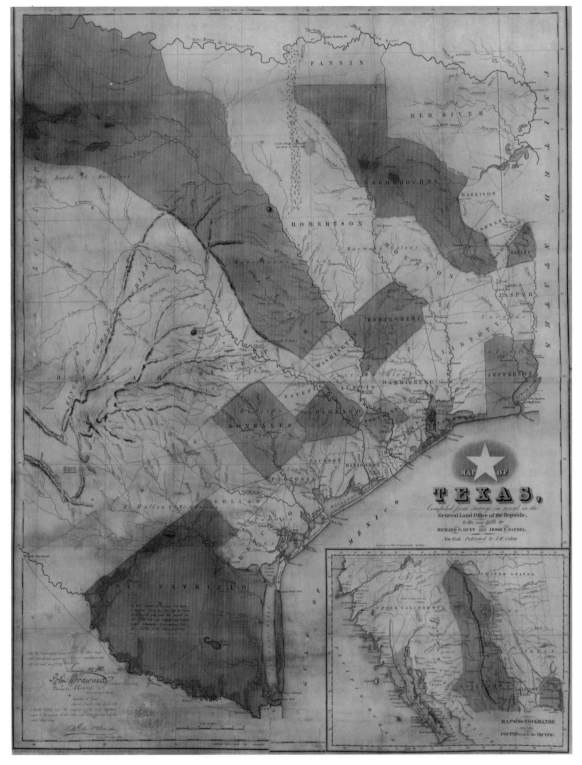

PLATE 20

Guide to the Republic of Texas . . . Accompanied by a New and Correct Map by Richard S. Hunt & Jesse F. Randel. Published by J. H. Colton in *Guide to the Republic of Texas;* New York, 1839. Lithograph by Sherman & Smith Stiles. 31.5" x 24" (Pocket Map).

This first travel guide for the new Republic of Texas became standard for American settlers. The map identifies 31 counties in color. The author claims the map was actually based on accurate surveys in the General Land Office. Newly founded Austin city is shown in central Texas along north bank of Colorado River. The inset shows Upper and Lower California.

treaty also included a finger of land extending all the way into southern Wyoming. These boundaries would be impossible for the Republic to secure; Comanche Indians controlled vast stretches of the territory, and Mexico still maintained a strong economic and military presence in Santa Fe, which the Texans now claimed. The Mexican government, in Santa Anna's absence, would soon repudiate the Treaties of Velasco, but even if it had been willing to acknowledge Texas independence, Mexico would never have agreed to a Texas that was double its historically recognized size (Plate 20).

Realizing the critical need for a permanent government that could command the respect of the Texas people and of foreign powers, Interim President Burnet called for early elections to choose a permanent government. Stephen F. Austin, who had inaugurated Anglo colonization in Texas and who for fifteen years had been the acknowledged leader of the colonists, soon announced his candidacy for president. Henry Smith, who as provisional governor had headed the failed state government from October 1835 till March 1836, also declared his candidacy. Three weeks before the election, however, Sam Houston entered the race, claiming—with some accuracy—that public opinion demanded his candidacy and that he was the only man who could unite the country. His claims were vindicated by the results: Houston won in a landslide, with 5,119 votes to Henry Smith's 743 and Stephen F. Austin's 587. Voters chose newcomer Mirabeau B. Lamar, who had distinguished himself in action at San Jacinto, as their vice-president. Texans also approved the constitution overwhelmingly and voted nearly unanimously to seek immediate annexation to the United States.

With a clear mandate, Houston moved swiftly to forge a government of national unity. He wisely brought his two defeated rivals for the presidency into the cabinet—Austin as secretary of state and Smith as treasury secretary. He also sought to heal old political divisions. To counterbalance Austin, who had led the so-called Peace Party before the war, Houston named as minister to the United States William H. Wharton, who had been a leading figure in the War Party faction. Hoping to bring the unruly army under control, he appointed Thomas J. Rusk, a trusted lieutenant who had the respect of the troops, as secretary of war.

Of the many pressing priorities facing the new government, none was more vital than establishing diplomatic relations with other countries, especially the United States. Houston had an ally in his old friend and political mentor Andrew Jackson, who was nearing the end of his term as president of the United States.

PLATE 23

Texas Compiled from the Latest and Best Authorities by Jeremiah Greenleaf. Published in *A New Universal Atlas*; Brattleboro, Vermont, 1842. Lithograph by G. R. French. 11" x 13".

This edition is based on Burr's classic 1833 Texas map. (Martin & Martin Plate 30). The western border of the Republic extends to the Great Bend and south along the Rio Grande. Counties and land grants are separately colored. Three tribes of "Apaches" are identified. Texas claims land north of the Arkansas River. The eastern boundary is moved twenty miles westward to the Sabine River at the 32nd parallel.

Jackson moved cautiously, rightly fearing that the recognition of slaveholding Texas would ignite a dangerous controversy over slavery, but on his last day in office in March 1837, Old Hickory finally appointed a chargé d'affaires to the Lone Star Republic. The United States thus became the first country to recognize Texas independence; it would be another two years before any European governments would establish diplomatic relations.

While Texans waited anxiously for the diplomatic wheels to turn, they had a nation to run. The Republic's finances were in a dire condition. The Texas Congress lacked the resources to pay government or military salaries. The Revolution had left the Republic saddled with a national debt of $1.25 million, which was growing larger with each passing day. Although the government levied various taxes on imports, property, and livestock, these measures produced little immediate revenue. Attempts to borrow money on favorable terms largely met with failure.

Houston adopted a program of fiscal austerity to place the Republic on a sound financial footing. In May 1837, he furloughed all but six hundred soldiers of the Texas army, agreeing to pay their way to New Orleans if they wished to leave for the United States or offering 1,280 acres of land apiece if they opted to remain in Texas. This would not only cut costs but would also address the problems that an undisciplined military posed to law and order. Houston further economized by seeking to negotiate peace with Texas Indians whenever possible. But despite his attempts to keep expenses low, the needs of the new government far outstripped the capacity of the economy to produce needed revenues. In June 1837 the government began printing paper money in the form of promissory notes, but this only resulted in inflation and a further weakening of the economy. When Houston left office in 1838, the national debt was approaching $2 million.

Leaders of the Republic recognized that the viability of Texas depended on attracting immigrants. Fortunately, the Republic possessed almost unlimited quantities of land, which

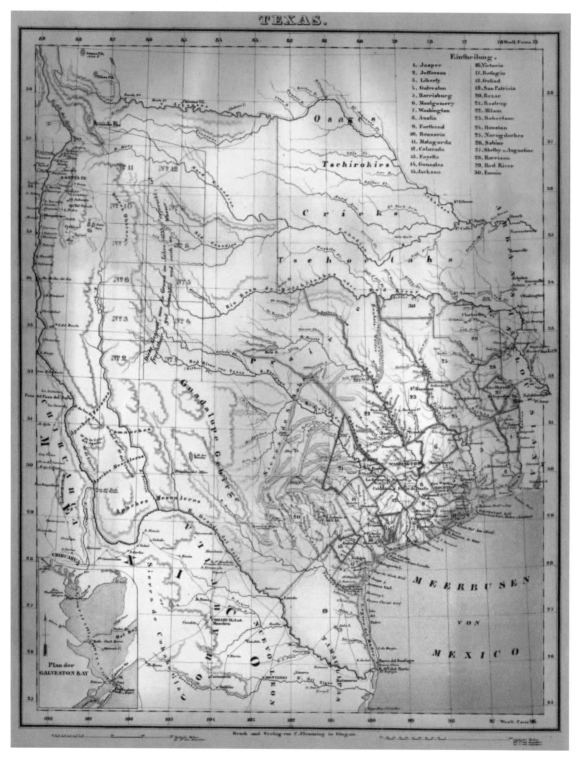

PLATE 24

Texas by Carl Fleming. Published in Glogau, Prussia, 1842. Drafting by Friedrich Handtke. 13.5" x 16.7".

This German edition of John Arrowsmith's 1841 Texas map shows the full extent of the Republic. (Martin & Martin Plate 32). The new nation's border claim extends to the upper Rio Grande River. Based on information from the General Land Office of Texas. The chart locates roadways, Indian tribes, and areas described as "excellent land" and "delightful country." The inset is of Galveston Bay, port of entry for most German immigrants, the intended audience of Fleming's edition. The map was published through 1853.

it was prepared to offer in generous quantities. During the Revolution the government had offered land grants to attract volunteers for the army; now Congress expanded the program. All veterans received varying amounts of land, depending on their service. Heads of families who had been in Texas at the time of independence received 4,600 acres; single men received about 1,500 acres. New arrivals could obtain 640 acres, on the condition that they live on the land for three years. Between 1836 and 1841, the government distributed close to 37 million acres. As a consequence of the lack of restrictions on the resale of land (or lax enforcement of such restrictions as existed) millions of acres fell into the hands of speculators (Plate 23).

The Constitution of 1836 prohibited presidents from succeeding themselves in office, and when Houston's term ended in 1838, Mirabeau B. Lamar won the presidency. Lamar's politics differed considerably from Houston's. The new president did not embrace the Jacksonian frugality of his predecessor, nor was he as enamored of annexation to the United States as was Houston. Lamar dreamed of an independent, even imperial, Texas that

would someday become a continental rival of the United States. He thus took a much more activist view of government than Houston did. Holding these attitudes, Lamar soon took Texas in some very different directions.

Nowhere were these differences more apparent than in Lamar's Indian policies. As president, Houston had pressed unsuccessfully for acceptance of a treaty he had negotiated in February 1836 for recognition of Cherokee claims in the Nacogdoches area. Lamar, in contrast, recommended that the senate reject the treaty. Soon after taking office, Lamar issued an ultimatum to the Cherokees, demanding that they relinquish their claims to land in Texas. When the Cherokees refused, he dispatched troops under the command of Rusk. Following their defeat at the Battle of the Neches in the summer of 1839, the remaining Cherokees and most of the other East Texas Indians were driven into Oklahoma.

Things were not so simple on the western frontier, where the Comanche Indians (immigrants to Texas in the mid-1700s) terrorized settlers with their brilliant horsemanship and fierce

PLATE 36

Karte von Indian Point od. Indianola, Karte der Stadt Neu Braunfels; Karte der Stadt Friedrichsburg; Hafen von Indian Point oder Indianola. (Carlshafen) by Johann David Sauerländer. Published by Carl von Solms-Braunfels in *Society for the Protection of German Immigrants in Texas* (author) *Instructions for German Emigrants to Texas, together with the newest Map of this State according to the boundaries determined by Congressional Resolution of September 1850.* Self-published by the Society's Committee: Weisbaden. 1850-1851. 23" x 17".

This provides the first plans for four cities settled by German emigrants: Indian Point; New Braunfels, Fredricksburg, and Indianola. It shows advanced planning by developers.

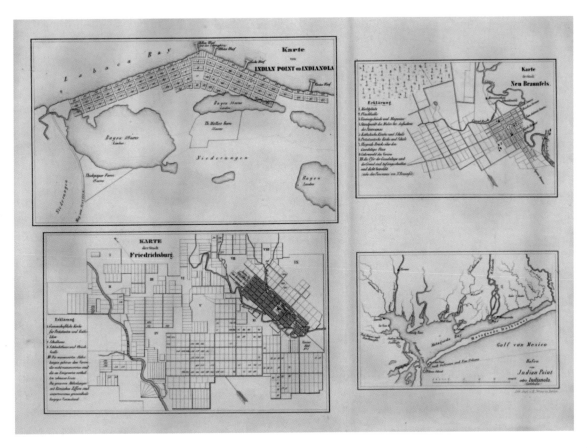

warrior code. In San Antonio in March 1840, thirty-five Co-manche and seven Texans were killed when negotiations for the release of captives broke down and a riot ensued. Following this incident (known by Texans as the "Council House Fight"), war with the Comanche raged on the frontier for a year. In August 1840, a thousand warriors plundered the towns of Victoria and Linnville before being dealt a decisive defeat by Rangers under the command of Ben McCulloch. After Sam Houston returned to office later that year, he managed to calm the situation, and the frontier returned to its usual pattern of occasional hit-and-run raids by the Indians and reprisals by the Rangers.

Lamar's management of the economy also contrasted sharply with Houston's. The new president resorted to issuing paper money, backed only by the presumed good faith and non-existent credit of the Republic. Not surprisingly, the money depreciated rapidly, to a point in 1841 where a Texas dollar was worth only about fifteen cents. Despite such setbacks, Lamar dramatically increased spending. Military spending alone—much of it spent in the aforementioned campaigns against Indians—cost some $2.5 million. Another ill-conceived policy, at least in terms of short-term fiscal considerations, was the decision in 1839 to move the Republic's capital from Houston to the site on the Colorado River that would later be named Austin. Although Lamar liked its central location as a future crossroads of trade, it was also isolated, vulnerable to attack, and lacking in ameni-ties. Expenses for such items as public buildings, travel for legis-lators, and mail service rose accordingly.

Lamar's imperial dreams also led him pursue other unwise projects. In 1841 he raised and outfitted a military and trade excursion into New Mexico, with the goal of incorporating Santa Fe and its lucrative transcontinental trade into Texas. The result-ing Santa Fe Expedition, consisting of 270 men commanded by Colonel Hugh McLeod, ventured into the uncharted wilds of West Texas, only to be plagued by heat, hunger, and Indians before being captured by Mexican troops. The captives were marched overland to Mexico City, where they were held until 1842 (Plate 24).

Efforts to populate Texas continued under the Lamar administration. Recognizing the inadequacy of existing policies, in 1840 Congress reestablished the empresario system, which had been so instrumental in attracting immigrants during the Mexican era. Under its provisions, several colonies were established. The most successful of these, the Peters Colony, was founded in 1841 in the Dallas area and ultimately attracted over 10,000

PLATE 27

Karte von Texas . . . general land Office de Republic . . . by Prince Carl von Solms Braunfels [Society for the Protection of German Emi-grants to Texas]. Published by J. D. Sauerländer's Verlag in Prince Carl von Solms-Braunfels *Texas.* Frankfurt am Main, 1845. Lithography by Geogr. & Lith. Anstalt von Eduard Foltz-Ebertle. 20.5" x 17".

This depiction of the Republic of Texas shows land grants held by the "Adelsverein." The map was part of an immigration guidebook issued by the organizers of the Society for German Emigration to Texas, by Carl Solms-Braunfels. Prince Carl describes Texans as "self-opinionated and boastful, unpleasing in their social dealings, and very dirty in their man-ners and habits." Insets are of Mexico and the Gulf Coast. The German Society bought an interest in the Texas Colonization contract from Henry Fisher and Burchard Miller in 1842. Prince Solms-Braunfels in 1845 purchased the Comal tract (site of New Braunfels), where 700 German emigrants were soon directed.

settlers, mostly from the Ohio Valley. In the Hill Country west of San Antonio and Austin, a minor German nobleman named Prince Carl of Solms-Braunfels established a colony that eventu-ally numbered some 7,000, including the towns of New Braunfels and Fredericksburg (Plate 36). To the south of the German set-tlements, Henri Castro received a contract to introduce Alsatian immigrants to an area southwest of San Antonio. Eventually some 2,000 colonists settled the area around Castroville (Plate 27). These settlers gave Texas a multi-ethnic flavor that is still evident today.

Sam Houston won a second term as president in 1841, in-heriting a $7 million national debt and a country still threatened by internal disorder and external enemies. Mexico continued to threaten to retake what it viewed as a renegade province, and in February 1842 a Mexican force commanded by General Rafael Vásquez crossed the Rio Grande and occupied San Antonio. The invasion was actually little more than a show of force—Vásquez stayed only two days—but in September a second force under General Adrián Woll again took San Antonio. This occupation was also brief, but under heavy public pressure, Houston ordered General Alexander Somervell to lead a volunteer expedition of about 750 men toward the Rio Grande to pursue Woll and to guard against future incursions. Although Sommervell ordered a halt at the Rio Grande, some three hundred of his men decided to cross into Mexico, bent mainly on plunder. At the village of Mier, on Christmas Day 1842, a Mexican army defeated the Texans in battle, capturing most of them. They subsequently made a break to escape, and 176 of them were recaptured. Afterward, the

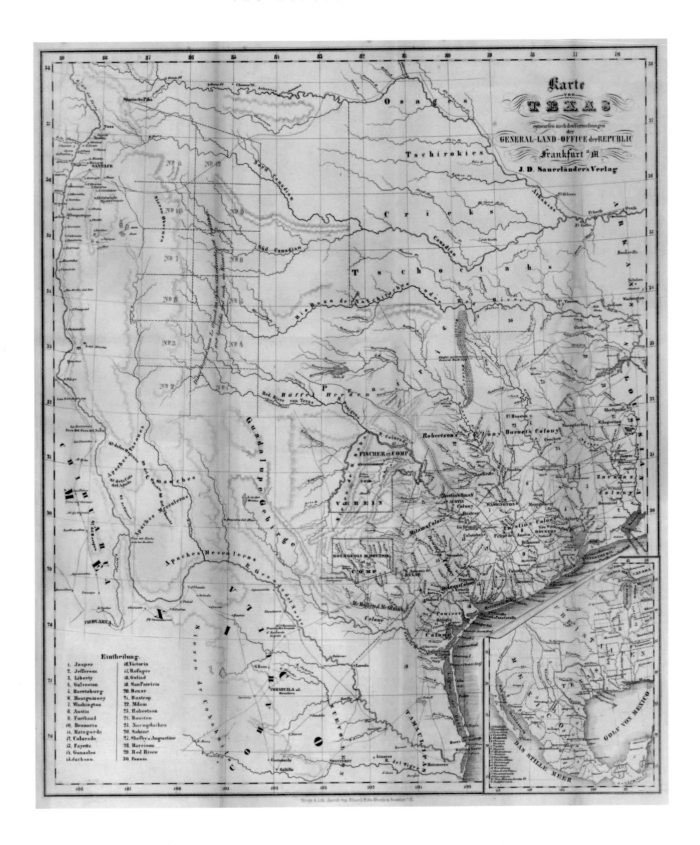

escapees were forced to draw from a pot containing 159 white beans and 17 black ones. Those unlucky enough to have drawn the black beans were executed by firing squad, and the rest were imprisoned in Mexico City. Coming so soon after the fiasco of the Santa Fe Expedition, the Mier Expedition further discredited the Texas government and underscored the necessity for annexation to the United States (Plate 30).

Despite these difficulties, Texas by the early years of the 1840s was showing signs of progress. Population growth was strong. Estimates place the population in 1846 at perhaps 150,000, a four-fold increase since the Revolution. Of these, some 30,000 were African-American slaves, a sign of the importance of the burgeoning cotton economy. Although Texas remained overwhelmingly rural and agricultural throughout the era of the Republic, towns did enjoy growth. Communities such as Gonzales,

Victoria, Brazoria, Velasco, Liberty, Nacogdoches, and Goliad, which pre-dated the Revolution, recovered from the war and in some cases enjoyed growth. New towns including Shelbyville, Richmond, La Grange, Columbus, Independence, and Clarksville came into being. Austin prospered as the seat of the Republic's government, and San Antonio gradually recovered from the destruction of the Revolution. Of particular importance were Houston and Galveston, which became major export points for Texas cotton and other agricultural products.

Transportation networks remained primitive and impeded the growth of the Republic. Roads were little more than cattle trails, which turned to quagmires when it rained. None of Texas' rivers were truly navigable, although shallow-draft steamboats did attempt to carry cotton from the interior to the coast on the larger rivers when water was high (Plate 28). The poverty of the

PLATE 30

Texas Nach den besten Quellen [Texas According to the Best Sources] by Carl C. F. Radefeld. Published by Joseph Meyer in *Grosser Handatlas*: Hildburghausen, 1846. 11.5" x 14.2".

This is a German edition of Emory's 1844 Map of Texas. The boundary does not include New Mexico because the Texan Santa Fe expedition failed to secure the new Republic's claim. The map shows the capital at Austin. Fort Alamo is next to San Antonio de Bexar. The explorations of Pike, Long, Freemont and Gregg are noted.

PLATE 28

Map of the Northwestern Part of Texas Received from the General Land Office in 1845 by Prince Carl von Solms-Braunfels. [Society for the Protection of German Emigrants to Texas]. Published by J. D. Sauerländer's Verlag in Prince Carl von Solms-Braunfels. *Texas.* Frankfurt am Main, 1845. Lithography by Maximilien Frommann, Darmstadt. 17" x 21".

The first published map of central Texas has a chart that identifies river boundaries of the Nueces, Colorado and Brazos rivers to the west, defining the German colony of Solms-Braunfels, and color outlines the German Emigration Company tracts in Bastrop and Travis counties. Twelve settlements are located: Austin, Bastrop, San Antonio, San Marcus, Gonzales, Comal Springs, Seguin, New Braunfels, Castroville, Fredericksburg, San Saba and Enchanted Rocks.

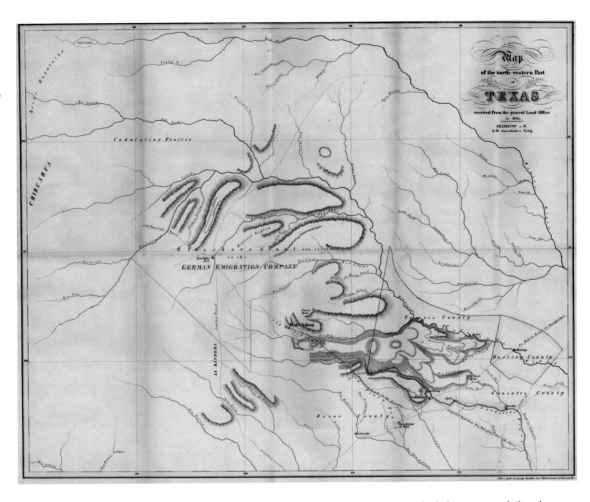

government and the overall lack of capital prevented the dredging of rivers and harbors.

Education never moved beyond an embryonic state. Lamar in 1839 and 1840 signed bills setting aside 17,714 acres of land in each county for the establishment of a primary school. Congress also set aside public lands whose proceeds were to be used for the establishment of two colleges. But nothing concrete ever came from these acts during the lifespan of the Republic: the lands yielded little or no income, and it would be decades before there would be a functioning system of public education in Texas. In its absence, education was left almost entirely in private hands. Affluent families generally sent their children to private schools back east or hired tutors. During its ten-year existence, the Republic granted charters to eight universities, seven colleges, ten academies, and four institutes—all of them sponsored by religious bodies. These included Rutersville College (1840), Wesleyan College (1844), and Baylor University (1845).

Although Texas was enjoying slightly more stability by 1843, most Texans continued to believe that annexation held the key to long-term security and prosperity. During his second presidential term, Sam Houston worked to create closer ties with France and especially with Great Britain. In part, he hoped that the European powers would be helpful in swaying Mexico to recognize Texas and thus end the Mexican military threat. But he also cannily courted these governments in hopes of persuading the United States to reconsider annexing Texas, lest America's European rivals gain a sphere of influence on the United States' western border. Expansionist sentiment was strong in the United States, and the spirit of Manifest Destiny had convinced many Americans that annexation was both desirable and inevitable. But neither the Jackson nor the Van Buren administrations had pursued annexation with much vigor, owing largely to the influence of the growing antislavery bloc in Congress and a fear that adding another vast slave state would weaken the bonds that held

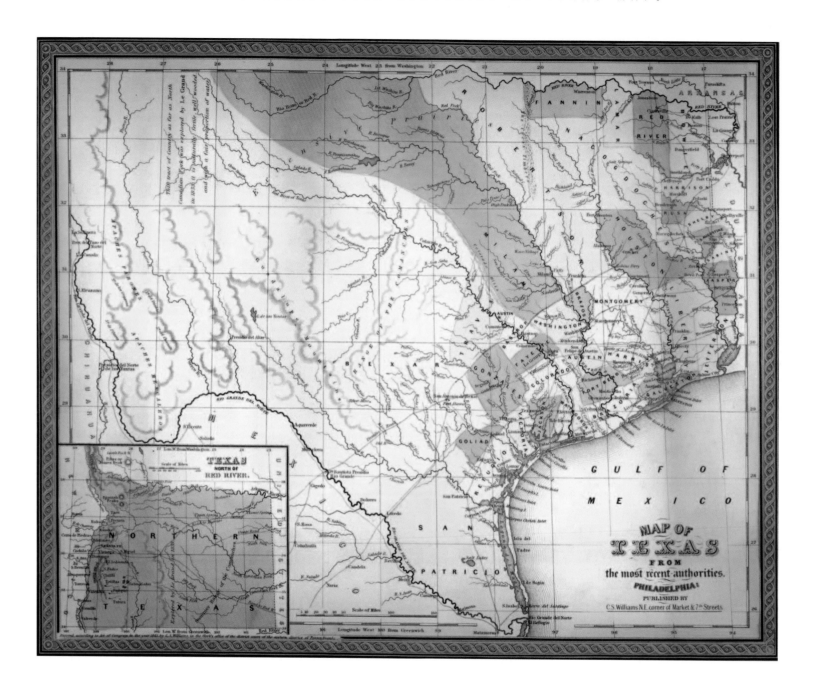

PLATE 26

Map of Texas from the most recent Authorities by C. S. Williams. Published in Philadelphia, 1845. Lithograph by J. H. Young. 12" x 1'.5".

This is the first edition of Samuel A. Mitchell's county map of the State of Texas, published in his New Universal Atlas. It shows thirty-four of the present-day 254 Texas counties. Robertson's County extends from the central part of the state north to the Red River. Bexar County includes all of the Panhandle and westward to the Rio Grande at the Big Bend. The chart contains Alexander le Grand's 1833 description of the fertile, wooded lands in Llano Estacado also noted on Arrowsmith's 1841 map. The inset shows northern Texas extending up to Santa Fe and present-day Colorado. The publishing company traces its history from Tanner through Mitchell to Thomas, Cowperthwait & Co.

PLATE 29

The State of Texas, 1836-1845 by David H. Burr. Published by J. H. Colton and R. S. Fisher in Pocket Map, *Map of Texas*; New York, 1846. Lithography by S. Stiles & Co. 17" x 21.3"

The original 1833 David Burr Texas map was the first large-scale map of the Republic. It was issued in four editions, the final in 1845. The 1846 Statehood edition was substantially updated, including new cartography for both the Gulf Coast and Rio Grande western border. Other additions included the towns of Corpus Christi, Aransas, Calhoun, Fort Brown and Fort Polk. The Burr map of Texas is second only to the Austin-Tanner map in importance. Only three known copies exist, all in public institutions.

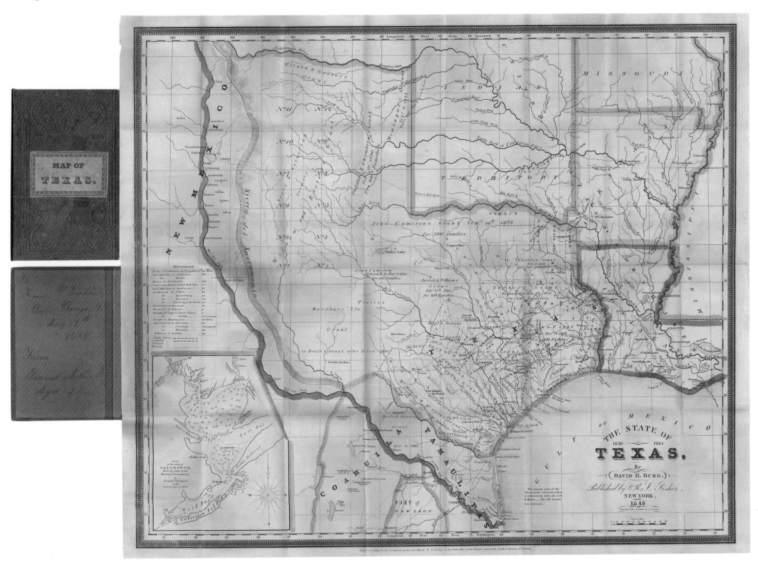

the Union together. Some Americans, particularly in the Whig political party, also feared that annexation would spark a war with Mexico (Plate 26).

In 1844 Texas negotiated a treaty of annexation with the United States, but the United States Senate rejected the treaty by a vote of 35-16. The Texan cause was not helped by the utterances of American Secretary of State John C. Calhoun, who had explicitly championed annexation on the grounds that it would expand and protect slavery. But matters took a favorable turn when the Democratic Party nominated a surprise candidate, Tennessee's James K. Polk, on a pro-annexation platform. Many saw Polk's victory in the 1844 election as a mandate for annexation. Among these was outgoing president John Tyler, who seized upon the maneuver of annexing Texas by means of a joint resolution of Congress, which required only simple majorities of both houses, rather than the two-thirds Senate vote required to ratify a treaty. Following an often-bitter debate, the bill passed, and on March 3, 1845, Tyler signed the bill. According to the terms of the act, Texas reserved the right to divide itself into as many as five states. More important, the new state retained ownership of its vast public lands as well as its enormous public debt. The western

border remained in dispute, with Texas continuing to claim the expansive boundaries that had been specified in the Treaties of Velasco (Plate 29).

The incoming president James Polk formally signed the Texas Admission Act on December 29, 1845. In a ceremony at the capitol in Austin on February 19, 1846, the last president of the Republic, Anson Jones, lowered the Lone Star Flag and raised the Stars and Stripes, declaring, "The Republic of Texas is no more."

SUGGESTED READINGS

Haley, James L. *Sam Houston.* Norman: University of Oklahoma Press, 2002.

Haynes, Sam. *Soldiers of Misfortune: The Somervell and Mier Expeditions.* Austin: University of Texas Press, 1990.

Hogan, William Ransom and Gregg Cantrell. *The Texas Republic: A Social and Economic History.* 1946; reprint ed., Austin: Texas State Historical Association, 2006.

Siegel, Stanley. *A Political History of the Texas Republic, 1836-1945.* Austin: University of Texas Press, 1956.

Uncharted Ground
Mapping West Texas, 1848-1861
Larry Francell

— ◆ —

The discovery of gold in California presented an immediate need for a national communication and a transcontinental transportation network. The period between the Mexican-American War and the Civil War would be one of intense scientific exploration of the American West. The United States Army took the lead in these endeavors, both keeping the lines of communication open and exploring the new domain.

The Spanish were familiar with the drainage systems of the Rio Grande and the Pecos River and were thus somewhat familiar with West Texas. However, at the beginning of the American period, the maps of the western part of Texas were devoid of detail (Plate 15). Although there had been great desire to take some of the Santa Fe trade away from Missouri, the Republic of Texas did nothing with its western domain. Missouri merchants controlled the main supply route to Santa Fe, and with this monopoly, they also controlled the lucrative Chihuahua City trade as well.

In an attempt to shorten the Missouri to Chihuahua City route, the first effort to explore West Texas in the American period was led by Dr. Henry Connelly, a Missouri trader. In 1839-1840, Connelly crossed the Red River and journeyed through the central part of the Republic to Horsehead Crossing on the Pecos. From there he and his men proceeded to Comanche Springs (Fort Stockton), Paisano Pass (Alpine), and down Alamito Creek to La Junta, the junction of the Rio Conchos and Rio Grande near the present community of Presidio. Their route then took them to Chihuahua City. However, this cut-off was not used until ten years later when the TransPecos section to Chihuahua City became part of the major trading route from San Antonio.

No other formal survey was undertaken until after the Mexican-American War, which generated much interest and information about the Southwest. With the Treaty of Guadalupe Hidalgo, 1848, and the Gadsden Purchase, 1853, Mexico ceded the entire Southwest. When gold was discovered in California in 1848 the necessity for roads to connect the East with California and to establish an exact boundary with Mexico touched off a flurry of exploration that in little more than a decade produced an immense wealth of information about the West.

The urgency for an all-weather route to California, a desire to tap the lucrative trade with northern Mexico through Chihuahua City, and the budding national interest in a transcontinental railroad route stimulated the exploration of West Texas. There were several allies aiding this endeavor. The first two were the powerful senators from Texas, Sam Houston and Thomas Rusk, both of whom advocated a transcontinental survey through Texas and supported a railroad through the state that would terminate in California (Plate 31).

The other primary assistance came from the Army Corps of Topographical Engineers. In 1838, at a most opportune time, science and politics would intertwine to provide a new approach to westward exploration. On July 5, the Topographical Engineers was created as a separate entity from the Corps of Engineers. An elite branch of the army with never more than thirty-six officers on the roster at any one time, the corps became the primary agent in the scientific mapping and study of the West for over twenty years. This handful of men laid out the national boundaries, surveyed wagon roads and potential railroad routes, charted major rivers, and generally provided the first systematic survey of the West.

Commanded in Washington by Colonel John James Abert, the corps was created during the rise of the expansionist political philosophy, Manifest Destiny. The basic tenet of this belief was that America, by divine right, was destined to expand and possess the lands west of the Mississippi to the Pacific and from Texas to Oregon. Some believed in a manifest right to lands much further north and south of those provinces.

In the West the corps became the direct instrument of American nationalism and Manifest Destiny. For the first time, professionally trained scientists and explorers took to the field to answer sophisticated questions that went beyond landmarks and trails. From its creation the corps would take the lead in western exploration directed towards three basic national concerns: first, a final geographic and political definition of the region; second,

a scientific inventory of the natural and human resources; and third, possible solutions to crucial transportation issues. Some of the best officers in the corps were stationed in Texas, under the local command of Brevet Lieutenant Colonel Joseph E. Johnston. His various subordinates included lieutenants William H.C. Whiting, William F. Smith, Francis Bryan, and Nathaniel Michler.

The first post-war expedition to enter the field was funded by commercial interests in San Antonio. These merchants enlisted John Coffee Hays and an escort of Texas Rangers under the command of Captain Samuel Highsmith to find a practical route to El Paso. The party of thirty-seven left San Antonio on August 27,

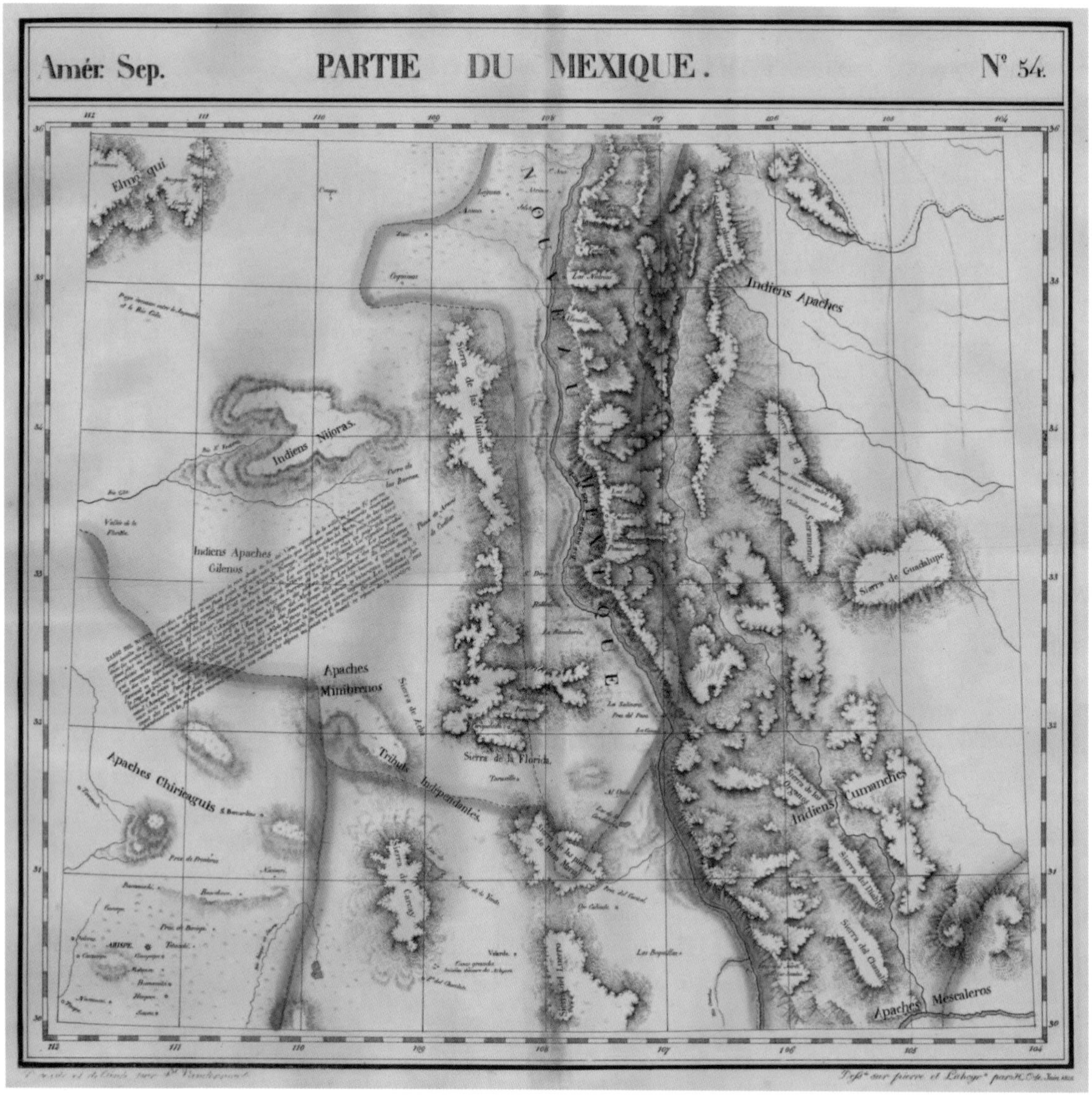

PLATE 15

Partie du Mexique No. 54 by Philippe Marie Vandermaelen. Published in *Atlas Universel de Géographie Physique, Politique, Statistique et Minéralogique*; Brussels, 1827. Lithography by Henri Ode. 18.5" x 20".

The Belgian map maker used a larger display for this series of American maps. Sadly, no new information was added. Map No. 54 presents southern New Mexico, northern Mexico and far western Texas. Mexican provinces are denoted by colored borders. "Passo del Norte," is described in detail, and local native American groups are boldly lettered. Mountains are misplaced, relying on the earlier geography of Humboldt (1811). It is more impressive than informative.

PLATE 31

Map of Gilliam's Travels in Mexico Including Texas and Part of the United States
by Thomas Sinclair. Published by Albert M. Gilliam in *Travels over the Table Lands and Cordilleras of Mexico during the years 1843-1844*; Philadelphia, 1846. 22" x 19".

Albert M. Gilliam, appointed U.S. Consul in San Francisco, wrote a diary of his western travels including two maps of the territory. The Texas portion of the map depicts two proposed western rail lines. One, styled " Pacific Rail Road to Napolian (sic)," passes Santa Fe to connect the Mississippi River to San Francisco. The second connects San Francisco to New Orleans. Neither was built because the geography was too rough. This is the first proposed railroad across Texas. Other than that novelty, the map is more often noted for its misinformation and out-of-date cartography.

1848, and returned three and a half months later, after encountering great difficulty. Close to starvation, the expedition made it only as far as Presidio, making no attempt to reach El Paso. Trying to traverse the harsh Big Bend country, both Hays and Highsmith reported that they had found a trail but in reality this project accomplished little. By the time Hays and Highsmith returned to San Antonio, the news of the discovery of gold in California reached the East. A southern all-weather route across the continent then became an imperative. As 1849 dawned gold

seekers began to arrive at the major population centers of Texas, ready to leave at the first opportunity. When cholera broke out, many started west before the end of winter and before a satisfactory trail could be surveyed.

Understanding the immediacy of the problem, Brevet Major General William Worth, the commanding officer in Texas, ordered Lieutenants William H.C. Whiting and William F. Smith to find a suitable route to El Paso. Departing from San Antonio on February 9, 1849, they were instructed to follow the Hays-

Highsmith trail as far as Presidio and then continue up the Rio Grande to El Paso. They were further instructed that if this route was not feasible they were to find a more practical way home. Crossing the San Saba near Fredericksburg, the expedition went three days without water before reaching the Live Oak Creek near the Pecos River.

Traveling through the Davis Mountains to Presidio, they followed the Rio Grande to El Paso. This outbound route was unsatisfactory. Hoping to find a better trail, especially one with more water, they returned by following the Rio Grande for only the first one hundred miles. At that point they marched due east for the Pecos. Following this watercourse south for sixty miles, they crossed to the Devils River, proceeded to its junction with the Rio Grande, and then went east to San Antonio via Fort Clark. Whiting and Smith reported that their return route was satisfactory, and it would soon be one of two major routes west through Texas.

While Whiting and Smith were in the field, another expedition was formed as a cooperative venture between commercial interests in Austin and the military. Austin merchants raised the funds necessary to send John S. "Rip" Ford to seek a route to El Paso, and General Worth sent Major Robert Neighbors to represent the military and federal interests. Neighbors was the Federal Indian Agent for Texas, and Ford, a former Texas Ranger, was publisher of the *Texas Democrat* in Austin. This party left Austin in March 1849 heading northward along the Colorado River to Brady's Creek, a tributary of the San Saba River. Striking west to the Concho River and Horsehead Crossing on the Pecos, they traversed the stark desert north of the Davis Mountains to El Paso. Not finding enough water to make this a practical road, the expedition returned by a more northerly route. Leaving El Paso they followed the general course of the present Texas-New Mexico border through the Guadalupe Mountains to the Pecos. Moving south along the river to Horsehead Crossing, they returned to Austin via Fredericksburg and San Antonio. Ford and Neighbors reported that the return route, with only minor improvements, would make an excellent wagon road.

During this same spring of 1849, Captain Randolph Marcy, Fifth Infantry, was dispatched by Colonel John Abert, commander of the Corps of Topographical Engineers, to survey a potential road from Fort Smith, Arkansas, to Santa Fe, New Mexico. In 1839 Josiah Gregg had traversed this same area as he sought a shorter trail for his Santa Fe trade. Marcy deemed this an excellent route, but it did not come into prominence (Plate 25).

PLATE 25

A Map of the Indian Territory, Northern Texas and New Mexico Showing the Great Western Prairies by Josiah Gregg and Sidney Morse and Samuel Breese. Published by Josiah Gregg in *Commerce of the Prairies* (NY: G. H. Langley, 1844) and *Atlas* (NY: Harper & Bros., 1845). 12.5" x 15".

The map was produced by the newly developed cerographic process, as indicated by the characteristic blue-line underlay. The map, considered a landmark, combines Humboldt's New Spain (1811); Major Long's notes from his first expedition (1840); and, J. C. Brown's survey of the Santa Fe Trail (1841), as corrected by Gregg's personal observations. The Staked Plains of Texas (Llano Estacado) are a predominate feature. The map became a cornerstone for the Santa Fe travelers to northern Texas.

General Worth died of cholera on May 7, 1849. His successor was Brevet Brigadier General William S. Harney. Even before Whiting and Smith and Ford and Neighbors returned, he had plans for a further survey of the two possible roads. Harney was more interested in the southern route and assigned Lieutenant Colonel Joseph Johnston to this project. When Whiting and Smith returned, Smith was ordered immediately to turn around and accompany Johnston. Johnston's party of engineers, escorted by a company of the First Infantry, was to proceed in conjunction with Major Jefferson Van Horne and his six companies of Third Infantry on their way to El Paso to establish a garrison. Also accompanying the expedition was a party of California-bound emigrants. The Topographical Engineers surveyed ahead of the main column, and Captain Samuel G. French of the Quartermaster Department commanded a detail assigned to make the necessary road improvements. Arriving without incident at El Paso on September 3, Johnston's party made only small changes to Whiting and Smith's original route. This trail would be known as the Lower, or Military, Road (Plate 37).

Harney assigned Lieutenant Francis T. Bryan to re-survey the northern route of Ford and Neighbors. On June 14, 1849, Bryan departed San Antonio westward to the Pecos. The party found easy passage all the way to El Paso, which was reached on July 29. The major problem with this potential route was a lack of water between the Concho and Pecos rivers. In his report Bryan suggested the possibility of digging wells in this area. Once in El Paso, Bryan joined Johnston and in October both officers and their men returned to San Antonio by this northern route. By this time a trail was well established and was soon known as the Upper Road. The final connection for the network of western

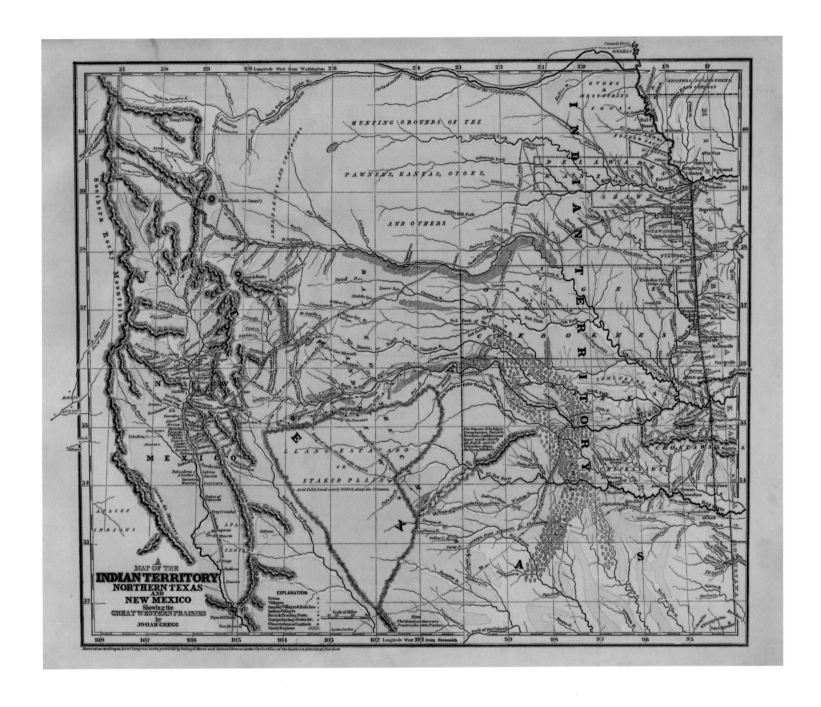

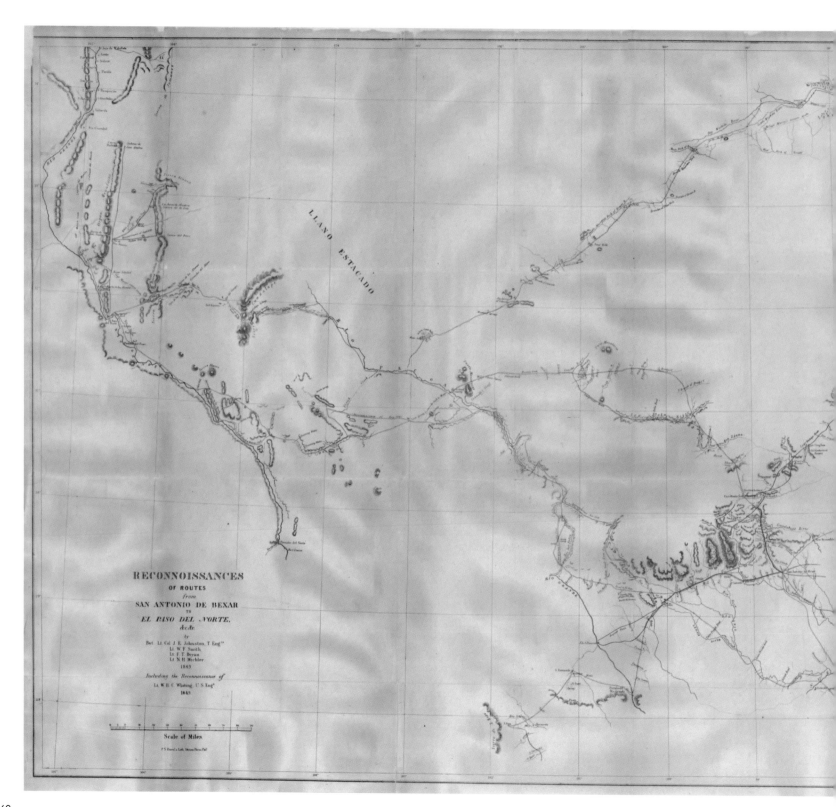

RECONNOISSANCES
OF ROUTES
from
SAN ANTONIO DE BEXAR
TO
EL PASO DEL NORTE,
&c.&c.
by
Bvt. Lt. Col. J. E. Johnston, T. Eng.rs
Lt. W. F. Smith,
Lt. F. T. Bryan
Lt. N. H. Michler
1849
Including the Reconnoissance of
Lt. W. H. C. Whiting, U. S. Eng.rs
1849

Scale of Miles

P. S. Duval Lith. Steam Press, Phil.

PLATE 37

Reconnoissances [sic.] of Routes from San Antonio de Béxar to El Paso del Norte by Lt. Col. Joseph E. Johnston Topographical Engineers. Published by the U.S. Government, Union Office in *Reports of the Secretary of War, with Reconnaissances of Routes from San Antonio to El Paso . . . also the Report of Capt. R. B. Marcy's Route from Fort Smith to Santa Fe . . . the Report of Lieut. J. H. Simpson of an Expedition into Navajo Country; and the Report of Lieut. W. H. C. Whiting's Reconnaissances of the Western Frontier of Texas*; Washington, D.C., 1850. Lithograph by P. S. Duval Lithographic Steam Press, Philadelphia. 24" x 36".

This military route is considered the cornerstone of West Texas maps. The surveys of Capt. Marcy, Lts. Whiting and Smith, the Emigrants Trail, the Comanche Trail, Old Spring Road, Connelly's Trail, and the Road to El Paso are all included. The two blue spots at the lower right are the Gulf of Mexico at Corpus Christi and Matagorda Bay. Only geography along the line of travel is given. The rest of the country's topography is left to the imagination.

roads across Texas was the establishment of a route from Fort Smith, Arkansas, that joined the Upper Road at the Pecos River. Captain Randolph Marcy of the Fifth Infantry and Lieutenant Nathaniel Michler were responsible for this survey.

Before the official surveys were completed, the Lower and Upper roads were in use by gold seekers on their way to California. Over 3,000 emigrants left from Texas in 1849, with many going by way of northern Mexico. Others followed the surveyors or attached themselves to military columns such as that of Johnston and Van Horne. Emigrants starting for the gold fields from Fort Smith or Dallas used the Upper Road. Those booking sea passage to the ports on the Gulf of Mexico and then traveling overland to San Antonio used the Lower Road, which also came into prominence as an alternative trade route to Chihuahua City (Plate 35).

Longer than the Santa Fe Trail, the Chihuahua Trail had the advantage of a port terminal on the Gulf of Mexico at Indianola on Matagorda Bay. This trail passed through San Antonio, where it joined the Lower Road as far as Comanche Springs (Fort Stockton), and then headed south to Presidio and Chihuahua City. More difficult than the Santa Fe Trail, this was still a profitable route for those strong enough to survive the hardships. At its height two hundred companies were engaged in this trade. Business along the Chihuahua Trail was interrupted by the Civil War but survived until the coming of the railroads.

The establishment of the Upper and Lower roads did not end the exploration of the TransPecos. On February 2, 1848, Nicholas Trist, on behalf of the United States, signed the Treaty of Guadalupe Hidalgo ending the war with Mexico. With the exception of three issues, numerous details related to the actual boundary between the two nations were left undetermined. San Diego was to be located north of any final boundary; both nations would have an outlet from the Rio Grande to the Gulf of Mexico; and the community of El Paso, currently the city of Juarez, would remain part of Mexico.

Used as the authority to determine a baseline for the boundary was John Disturnell's 1847 Map of the United States. This particular document was wrong in two significant areas. It located the Rio Grande two degrees of longitude too far west and the community of El Paso del Norte forty minutes of latitude too far north, or approximately thirty miles (Plate 16). The entire boundary in the Texas-New Mexico region was affected by this distortion, which created infinite controversy.

The treaty stipulated that each government was to appoint a commissioner and a surveyor who were to meet in San Diego

PLATE 35

Map of Mexico & California
by Julius Hutawa. Published by
Missouri Republican for Chambers and Knapp; St. Louis, 1849.
Lithograph by Julius Hutawa
Lithographer. 24" x 19".

*This is the second edition of a
map produced by the St. Louis
lithographer and printer who
specialized in providing city views
and maps for western travelers
and explorers. Originally the
map was issued as a supplement
to the St. Louis Missouri Republican in 1847. The California
geography is based on Frémont's
1845 survey. Exploration routes
of Smith, Gregg, Kearny, Cooke,
Lewis & Clark are noted. The
Oregon, Santa Fe and Old
Spanish trails and trading posts
are marked. This provides the
most detailed information for
the Santa Fe region.*

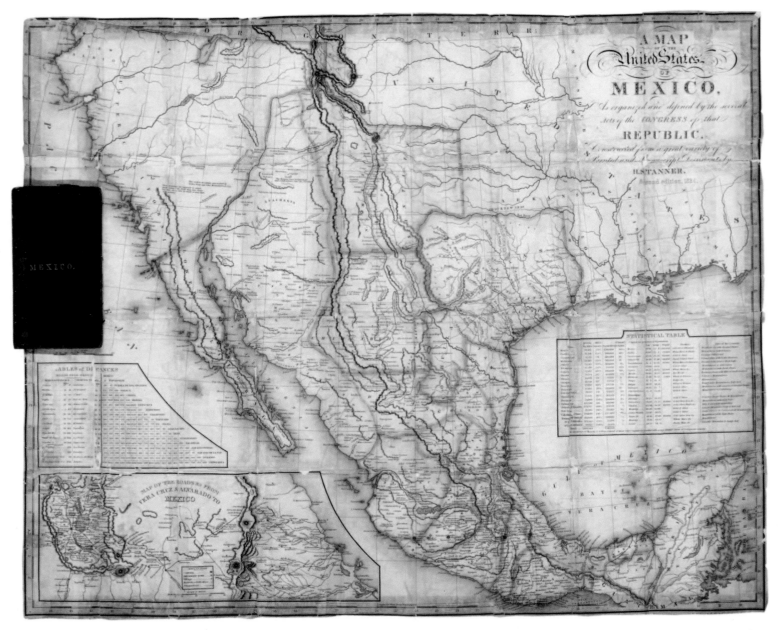

PLATE 16

**A Map of the United States of Mexico as Organized and Defined by . . .
Congress of that Republic** by Henry Schenck Tanner. Published by H.S.
Tanner in Pocket Map in Boards, "Mexico"; Philadelphia, 1834. 23" x 29".

*This large pocket map includes two insets ([1] Table of Distances [2]
Map of Roads from Vera Cruz to Mexico). With an April 2, 1832,
copyright but a 1834 edition, this is commonly considered the source*

*for John Disturnell's plagiarized "Treaty Map," used in 1846. Tanner's
original sources were Humboldt, Darby, and Pike. Five editions of
Tanner's popular map were issued between1825 and 1847. Early editions showed the boundary between Lower and Upper California 120
miles south of San Diego, an error corrected by the Treaty of Guadalupe
Hidalgo and the Gadsden Purchase. This edition used information from
the more famous Austin-Tanner map of Texas.*

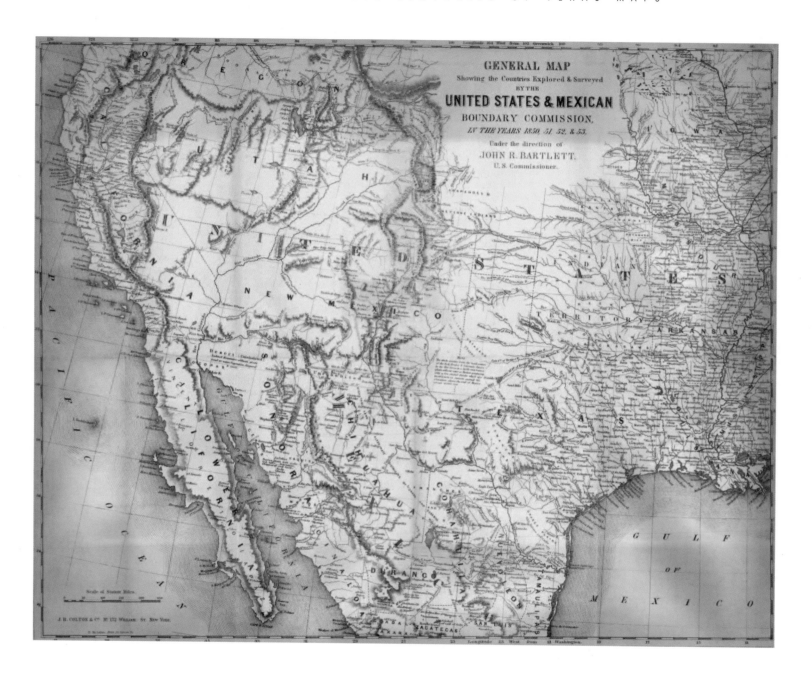

PLATE 38

General Map . . . United States and Mexican Boundary Commission . . .
Under Direction of John R. Bartlett, U. S. Comr. by J. H. Colton & Co.
Published by John Russell Bartlett for D. Appleton & Co. in *Personal*
Narrative of Explorations . . . In Texas; New York, 1854. Lithography
by D. McLellan, Print. 16" x 20".

The Treaty ending the Mexican American War provided for a Boundary
Commission to establish a new border. The commission first met on July
6, 1849. John Russell Bartlett replaced John C. Frémont as U. S. Com-
missioner after Frémont's election to the U. S. Senate in 1850 for the new
state of California. The so-called Disturnell Treaty Map was inaccurate
and renegotiations followed, resulting in a compromise that ultimately
led to the Gadsden Purchase of 1853. After a personality dispute,
William H. Emory joined the commission in 1851. He completed the
survey on October 15, 1855. The map shows Gill as the southern
international boundary.

within a year. The actual work began in July 1849 when United
States Commissioner John B. Weller of Ohio and surveyor An-
drew B. Gray of Texas met General Pedro Garcia Conde, the
Mexican commissioner, and surveyor Jose Salazar Ylarregui.
The army provided an escort under the command of Brevet Major
William H. Emory, assisted by Lieutenant Amiel Weeks Whipple.
Emory would be associated with the Boundary Commission
throughout the project and would be instrumental in ensuring
there was a final determination.

Weller was a Democrat at a time when Whigs controlled
Washington. In what was overt partisan politics, in June 1849
Weller was replaced and John C. Frémont appointed commis-
sioner. After several months of hesitation, Frémont declined the
job and accepted the position of United States Senator from
California. After almost seven months with the survey in limbo,
John Russell Bartlett replaced Weller. Bartlett was a Whig, a New
York bookstore owner, and founding member of the American
Ethnological Society. By this point the survey was essentially
complete from San Diego to the Rio Grande. Major Emory was
in charge of field operations, a position important to the success
of the enterprise.

The commissioners, Bartlett and Conde, met in El Paso in
December 1850. Their attempt to establish the boundary based
upon Disturnell's map seriously distorted the concepts of longitude and latitude. Realizing that Bartlett was about to cede a
strip of land that he considered important as "the great gateway
to the Pacific" and a potential transcontinental railroad route,
Andrew Gray, the official surveyor, refused to sign the offered
agreement. Major Emory was charged with setting things straight.

Technically working under Bartlett, who spent his time pursuing
his personal interests, Emory reorganized the survey for greater
efficiency and also worked well with his Mexican counterparts
(Plate 38).

In 1856 Emory published his final report. Yet the potential
remained for conflict with Mexico until 1854, when President
Millard Fillmore dispatched James Gadsden, a South Carolina
railroad promoter, to Mexico to negotiate with President Santa
Anna. The result was the Gadsden Purchase, $10 million for the
disputed lands west of El Paso and the valley of the Gila River.
The last major surveys in West Texas occurred in 1854. By this
time there was a compelling need to determine a railroad route
to the Pacific. However, as a result of sectionalism, Congress was
divided equally between free and slave states. Due to the expense
of such an undertaking the federal government would have to
subsidize the construction of only one route. Hence the selection
of this route created one of the most complicated political problems of the period.

In March 1853 Congress authorized the Pacific Railway
Surveys, expected to be the final solution to the problem. The
War Department, under Secretary Jefferson Davis, was to survey
all of the principal routes and determine scientifically and objec-
tively which would be the most economical and practical. In
1854 several parties consisting of engineers, surveyors, natural-
ists, astronomers, teamsters and military escorts took the field
for the greatest scientific endeavor of the age (Plate 39).

While all of this activity was designed to unravel the political knots of selecting a transcontinental route in the presence of
intense sectional interests, this was not to be. After a year's work
and a published report in thirteen volumes, the surveyors found
there were several practical routes. Thus the chance for a non-
political solution died in 1855. It would be eight more years
before a resolution would be found and then only after the South
withdrew from the debate by seceding from the Union.

In October 1853, as one of the four major surveys, Lieutenant John Pope was ordered to find a possible route along and
near the 32nd Parallel from Dona Ana, above El Paso, to Preston
on the Red River. Lieutenant John G. Parke was to make the
survey from California to Dona Ana. Not starting until February
1854, Pope had two main tasks. The first was to find a suitable
pass through the Guadalupe Mountains, and the second was to
seek out sources of water on the Llano Estacado. Since most of
the route was already known through the work of Bryan, Marcy,
and Michler, Pope encountered few problems. With the provision
that artesian wells could be developed to provide water, he found

PLATE 39

Map Illustrating the General Geological Features of the Country West of the Mississippi River by James Hall and J. P. Lesley. Published by U.S. Government Printing Office to Congress, in William H. Emory, *Report on the United States and Mexican Boundary Servey . . .* Washington, D.C. Lithography by Sarony, Major & Knapp, New York; Draftsman, Thomas Jekyll.

22" x 23".

This early western geological map was compiled from notes by William H. Emory. Emory's cartography of West Texas while he served as the supervisor of the U.S. Mexican Boundary Commission (1848-1853) provided the scientific information.

PLATE 52

Map Accompanying Reports of the Artesian Wells Investigation . . . Western Texas Public Land Strips and Part of New Mexico and Indian Territory by Richard J. Hinton, Special Agent in Charge. Published by U.S. Government, Washington, D.C., 1890. Lithography by the Norris Peters Co., Photo-Litho., Washington, D.C., 1890. 35" x 29".

Groundwater trapped below rock strata was essential to the development of the arid western territories. Well sites were investigated beginning in the 1880s. The Trans-Pecos Basin proved fruitful, producing flowing wells in Fort Stockton and El Paso. Artesian irrigation added agricultural value to what was previously only open range. Artesian wells became the subject of state regulation under the General Irrigation Act of 1913. Artesian wells were also necessary to the transcontinental rail lines in the dry southwestern portion of the United States.

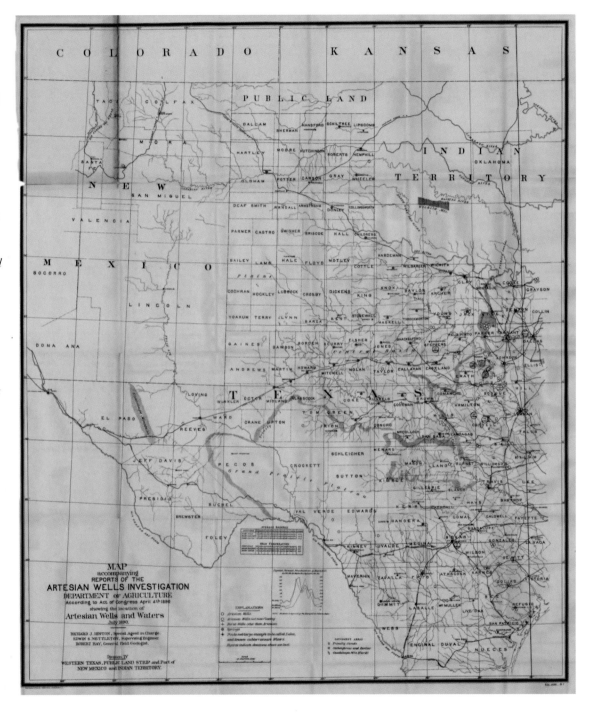

a route acceptable for a railroad. The next year Pope was sent back to the Staked Plains to dig wells. He made several attempts, but his equipment was inadequate and the mission was deemed a failure (Plate 52).

A second major survey, along the 36th Parallel, was assigned to Lieutenant Amiel Weeks Whipple. Starting at Fort Smith, this expedition essentially reworked ground that had already been traversed by both Josiah Gregg and Captain Randolph Marcy. Whipple also considered this an acceptable route for a transcontinental railroad.

The needs to define the national border with Mexico, find suitable roads west during the Gold Rush, and seek a practical transcontinental railroad route brought some of the best and the brightest men in the army to Texas in the decades between the Mexican-American War and the Civil War. Their combined work produced the first true geographic picture of Texas at a time when this information was vital to the national interest. Most of these men went on to major command responsibilities for either the North of the South during the Civil War.

SUGGESTED READINGS

Goetzmann, William. *Army Exploration in the American West.* Austin: Texas State Historical Association, 2000.

_____. *Exploration and Empire: The Explorer and Scientist in the Winning of the American West.* Austin: Texas State Historical Association, 2000.

Morris, John Miller. *El Llano Estacado: Exploration and Imagination on the High Plains of Texas.* Austin: Texas State Historical Association, 1997.

Rebert, Paula. *La Gran Linea: Mapping the United States – Mexico Boundary, 1849-1857.* Austin: University of Texas Press, 2001.

Viola, Herman. *Exploring the West.* Washington: Smithsonian Books, 1987.

ANNEXATION TO SECESSION

ROBERT MAYBERRY JR.

The funeral for the Republic of Texas took place on February 19, 1846. Held on the steps of the old capitol in Austin, it was attended by large crowds and officials of both the old government and the new. Anson Jones, the Republic's last president, delivered the eulogy and then hailed the birth of a new creation—the sovereign State of Texas. Flanked by a portrait of Stephen F. Austin and the three Mexican battle flags taken at San Jacinto, he proclaimed: "The Republic of Texas is no more." As Jones himself lowered the Lone Star flag for what all believed would be the last time, heads bowed amidst abundant tears. Then he ran up the Stars and Stripes in its place while cannon fired and as the brisk breeze snapped the bright banner into full view, "cheer after cheer rent the air." After nearly ten years of strife and uncertainty, the people of Texas finally had achieved their fondest desire—to have Texas added to the map of the United States. Despite the elation, unresolved issues clouded that happy day. Among these was how far the Texas borders extended. This most fundamental of questions led almost immediately to renewed conflict with an old enemy, Mexico.

Mexicans would not give up their claim to Texas without a fight, but it took the boundary question to provoke it. After San Jacinto, the victors had coerced captive General Antonio López de Santa Anna into signing the Treaties of Velasco, which asserted that the Rio Grande constituted the southern and western borders of Texas. Even by the rapacious standards of frontier Americans, this was an astonishing claim. Historically, neither the borders of Spanish nor Mexican Tejas had ever extended beyond the Nueces River, let alone to the banks of the Rio Grande. The vast territory Texans now claimed included the thoroughly Hispanic towns of El Paso, Albuquerque, and Santa Fe, most of New Mexico, and parts of present-day Colorado. In the unlikely

event Mexicans were to recognize the Republic of Texas, they would never accept the Rio Grande as the border.

Yet Texans claimed every right to the disputed land. Throughout the brief history of the Republic, armed bands, both with and without official endorsement, periodically descended on the Mexican settlements along the Rio Grande in quest of plunder and with an eye toward seizing control of the region from the Mexico City government. In 1841 Mirabeau B. Lamar, Texas' expansionist second president, launched an unsuccessful military expedition to Santa Fe aimed at asserting sovereignty.

James K. Polk, the quintessential Manifest Destiny president, believed the natural border of the United State should be the Pacific Ocean. The Texas claim was key to achieving his aims, and he was determined to press it. At first the president had tried to secure this goal through negotiation and purchase, but Mexico predictably rejected his overtures. Polk calculated that if money did not talk, American cannons could. As soon as it became clear that Texas would accept annexation, he sent General Zachary Taylor and a strong force of United States troops to Corpus Christi at the mouth of the Nueces. When further negotiations failed, Polk ordered the army to the Rio Grande. This provocation proved too much for Mexican officials, who dispatched their own formidable force north of the river. After a series of skirmishes and major battles at Palo Alto and Resaca de la Palma early in May 1846, the defeated Mexican army reeled back across the Rio Grande. Taylor then led his troops into northern Mexico, captured Monterrey, and thwarted a determined Mexican offensive at the Battle of Buena Vista. Meanwhile, General Winfield Scott commanded an invading army that fought its way from Vera Cruz to Mexico City, entering in triumph on September 13, 1847. While the major fighting had raged, smaller

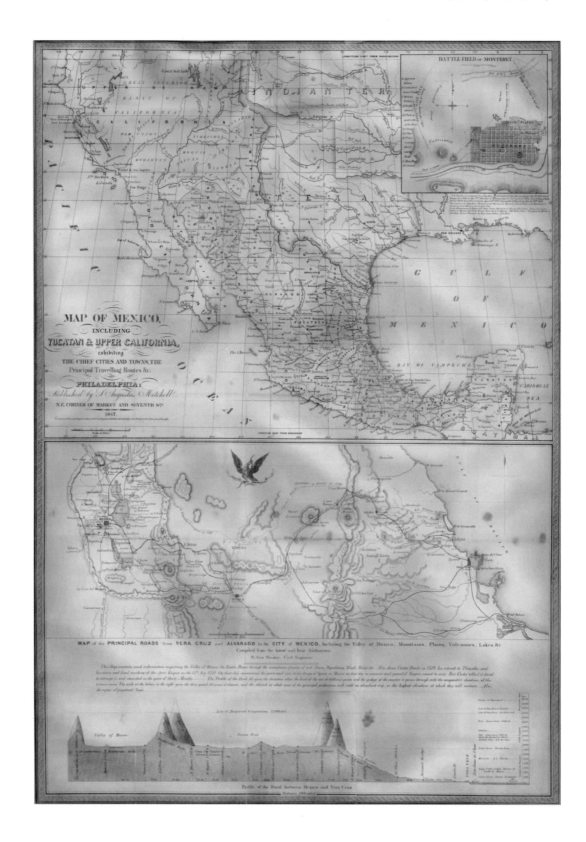

PLATE 32

Map of Mexico Including Yucatan & Upper California by Samuel Augustus Mitchell; 1847. 33" x 24".

This map chronicles victories by U. S. Forces in Mexico. The accompanying chart shows the full extent of the Republic of Texas. Insets include the Battle of Monterrey and route between Vera Cruz and Mexico City. All Mexican states are identified by color. The war began after the annexation of Texas to the United States. Two primary issues involved the Rio Grande and northern California borders. American troops also entered that country through Santa Fe. The Mexican coast was under U. S. Naval blockade. The capital of Mexico City was occupied in the fall of 1847. The Treaty of February 1848 secured the U. S. possession of California, Arizona, New Mexico and the Rio Grande border for Texas.

United States Army units had seized New Mexico and California (Plate 32).

Texans played an important role in the war but not a major one. Of volunteer troops, only Louisiana had mustered more than the 5,000 raised in the Lone Star State. The most conspicuous Texans were those that Texas Ranger captains Samuel H. Walker and Jack Hays led. These mounted hellions served as scouts and skirmishers who were often dispatched to counter Mexican raiders and guerilla bands. They nursed an implacable hatred for Mexicans nourished by raw memories of the Alamo, Fannin's demise, the Dawson Massacre, and dozens of vicious skirmishes during the days of the revolution and Republic. And the feeling was mutual. To Mexicans who accused them of all manner of atrocity and outrage, they were _Los Diablos Tejanos_. The Mexican War was one of the first that newspaper correspondents extensively covered, and for the gentlemen of the fifth estate the "Texan devils" made very good copy. An eyewitness described them as men "in buckskin shirts, black with grease and blood . . . all armed with Revolvers and huge Bowie Knives." The Mexican War may have been a victory primarily of the regular army, but Texans' well-known reputation for being colorful, fierce, and resilient citizen-soldiers was powerfully enhanced.

In February 1848 a humiliated Mexico signed the Treaty of Guadalupe Hidalgo, forfeiting New Mexico, California, and most of the present-day American Southwest, in addition to relinquishing all claims to Texas and acknowledging the border as the Rio Grande. In return for the Mexican Cession the United States paid $15 million.

While the Mexican War resolved one boundary problem, a more serious one soon attracted national attention: how to draw the map of American slavery. Northern politicians, opposed to

adding another slave state, had delayed Texas admission to the union nearly a decade. Until the Mexican War, the 1820 Missouri Compromise had assured parity between the free and slave states. Free soil advocates now demanded that the newly acquired territory be closed to slavery (Plate 33).

Texans naturally assumed victory had secured not just the southern border but its claim to the expansive western boundary as well. But residents of New Mexico, who included Yankee traders as well as Mexicans—a population unlikely to be favorably disposed toward Texans—fiercely opposed acquisition by the Lone Star State and instead petitioned the United States for territorial status. In the meantime, in 1848 the Austin government dispatched officials to organize Santa Fe County as part of Texas. United States military commanders in the area, however, rebuffed the Texans, and newly-elected President Zachary Taylor supported his officers. After a second commissioner failed to establish jurisdiction in early 1850, many outraged Texans advocated seizing New Mexico by force. A crisis loomed when President Millard Fillmore, who had succeeded the now-deceased Taylor, threatened to send troops to the region if Texans asserted their position by arms (Plate 34).

This saber-rattling echoed throughout the nation during the fateful year of 1850 as southerners began to threaten secession over this and a variety of unrelated issues involving slavery and the Mexican Cession. Reasonable men, however, still had influence in Washington, and the Senate passed a series of bills collectively known as the Compromise of 1850 that postponed an ultimate showdown for more than a decade.

One of these measures drew the final outline of the Lone Star State. The United States government agreed to "purchase" the disputed western lands for $10 million, which would be used to repay debts from the days of the Republic. Many "imperialist" Texans resented the alienation of so much land, but realism prevailed. In a special election Lone Star voters approved the settlement by a two-to-one margin.

With Texas' newly-defined borders secure, southerners found this beckoning "Eden" particularly attractive. With no frontiers to cross, no natural barriers, and no cultural divides, the land rush was on. In the decade after 1850 the population of Texas tripled. More than ninety percent of the immigrants shared a common southern Anglo-Celtic heritage, which assured that their traditional system of agriculture and animal husbandry would dominate the state's antebellum economy. Thus, three-fourths of the population comprised farmer-herders, who grew corn for personal consumption, raised a little cotton or wheat to sell, and fed cattle and hogs on the open range. As one settler later recalled, we were "by birth, blood and habitation, Southern people."

Yet, regional differences in crop selection, labor characteristics, and land usage were evident. Settlers from the deep South tended to cluster in the rich bottom lands of East Texas, where the more prosperous among them reproduced the slave-dependent, cotton-producing plantation economy they had left behind. In North Texas, except along the Red River, few antebellum settlers embraced large-scale cotton production, so slavery was never a dominant feature there. Instead, North Texans, who chiefly hailed from the upper and middle South, grew mostly wheat, which they sold to army posts in Indian Territory to supplement their herding. Migrants from all parts of the South met and melded in Central Texas, where they constituted a mixed bag of small farmers, herders, and planters. In South Texas below the Nueces River, southern open-range herding traditions fused with Latin vaquero technology to produce the beginnings of cowboy culture and the cattle industry.

Nevertheless, much of the state's economy depended on slave labor, and the "peculiar institution" remained firmly entrenched. By 1860, one in five Texans owned at least one slave, while slaves made up thirty percent of the total population. In Texas, however, slavery had reached its natural geographic limit; slaveholders lived almost exclusively east of the ninety-eighth meridian and to the north of the mouth of the Nueces River. Nevertheless, slaveholders represented a powerful force in Texas politics.

While newcomers were integrating uneventfully around established Texas communities, a bloody clash of cultures ensued on the western border of settlement. By the end of the Republic, Texans had driven most Indians out of the eastern part of the state, but the fierce mounted bands of the plains, like the Comanche and Kiowa, continued to threaten western expansion. Texas frontiersmen were determined to kill or remove any tribes that remained in the state, while Indian raiders viewed white settlements as legitimate targets.

Beginning in 1849, the United States Army constructed a line of eight posts from Fort Worth to Fort Duncan on the Rio Grande to protect the growing population along the frontier. But so rapid was the influx of newcomers that within two years the settlement line was a hundred miles farther west. This required building a new defensive line from Young County in the north to fifty miles west of Eagle Pass in the south. Despite these efforts

PLATE 33

The United States of America, the British Provinces, Mexico and the West Indies and Central America by George W. Colton; 1849. Lithograph by John W. Atwood. 50" x 57".

One of the last maps to show the Republic of Texas with its full stovepipe panhandle shape extending up to Wyoming, this edition responded to the public's excitement over gold in California. Frémont's California-Oregon cartography was used for the West. Steamship Trans-Atlantic crossings (inset) add information for would-be gold prospectors from Europe. The bold American eagle cartouche symbolizes completion of "Manifest Destiny." The map is surrounded along the border with public buildings and western views. The map was issued through 1854.

the army was never particularly successful in stemming the tide of Indian incursions. Not until the 1870s would Texans lay full claim to the western reaches of the state.

With most of the immediate issues of the day uneasily addressed, Texans turned their attention to developing the public institutions that a state with a rapidly increasing population and a building economy would require. With money received from a number of agreements with the federal government and using funds from the sale of state lands, Texans set about establishing public schools, constructing public buildings, and promoting railroad construction. But such efforts fell short of fruition. The clouds that had loomed in 1850 never quite receded from the horizon and in 1854 began to billow again. This time there would be no avoiding the storm.

Trouble began anew when Illinois Senator Stephen A. Douglas proposed a bill that would organize land west of Missouri in order to construct a railroad from Chicago to the Pacific. The Kansas-Nebraska Act, as it was called, dodged the issue of slavery by proposing "popular sovereignty" for any states formed from these lands. That is "all questions pertaining to slavery . . . are to be left to people residing therein." Southerners at last saw an opportunity to gain a slave state.

What they got instead was a dress rehearsal for civil war. Groups sprang up in both North and South whose purpose was to populate the Kansas territory with like-minded settlers. Pro-slavery "border ruffians" from Missouri were the first on the scene and were soon joined by determined immigrants from other southern states. Typical of these was John W. Whitfield, a Mexican War veteran and staunch pro-slavery man from Tennessee. In 1854 he moved to Kansas with his family, livestock, and possessions, hoping to make the territory his home and strike a blow for southern rights. Such men were soon outnumbered by

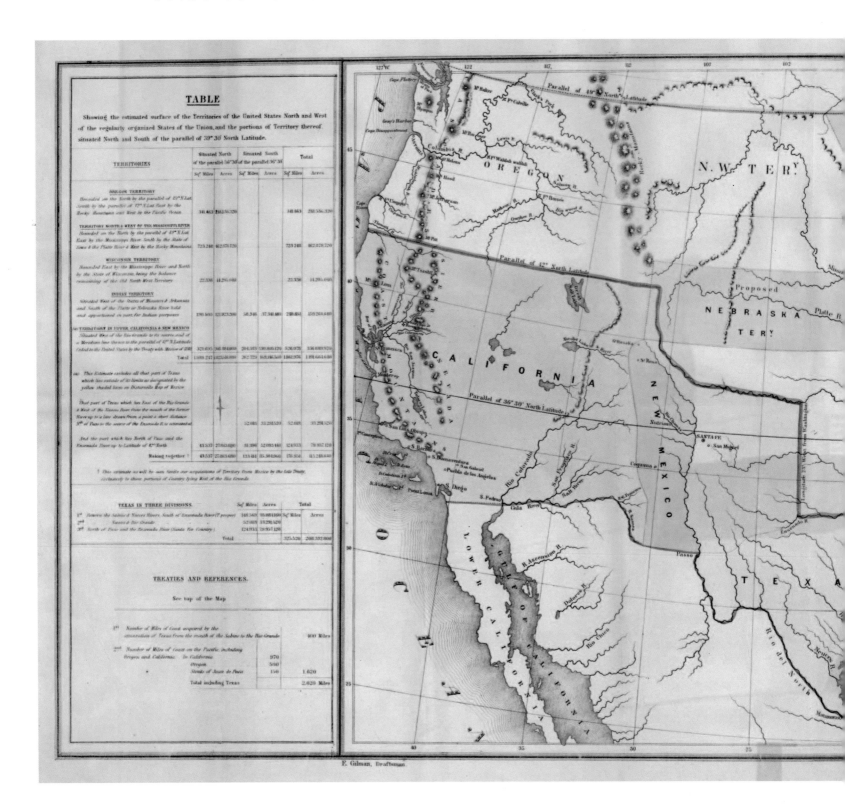

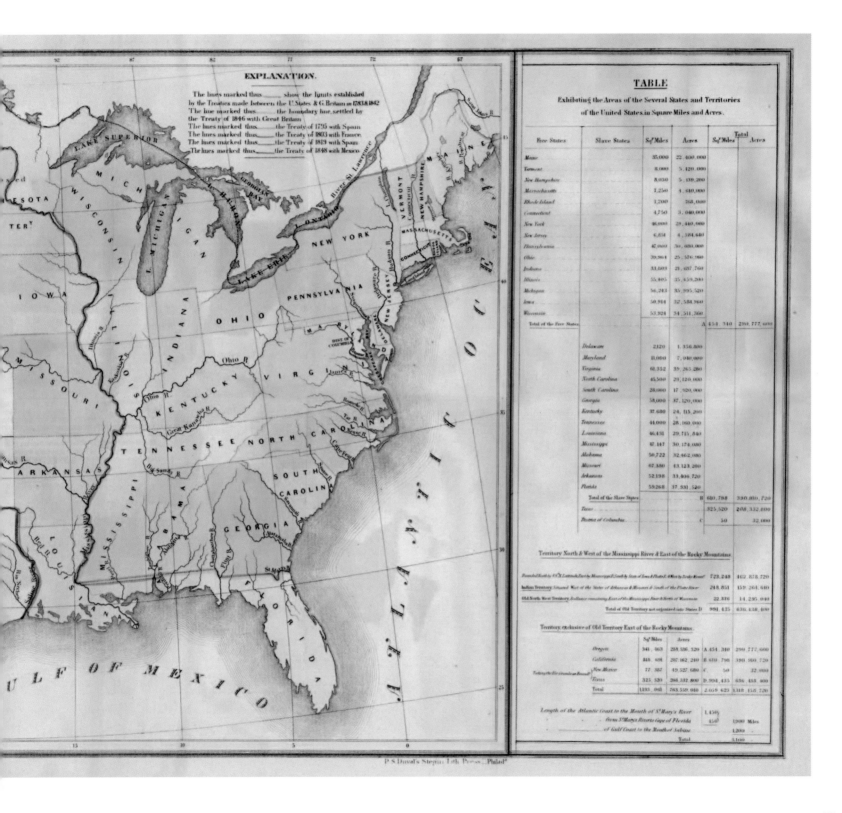

P.S. Duval's Steam Lith Press., Philad.

PLATE 34

[Lines of Treaty Map of United States Between 1783 and 1848] by E.
Gilman. Published by the U. S. Government; Philadelphia, 1848. Lithograph by P. S. Duval's Steam Lith. Press. 14" x 34".

*The map provides estimated surface territories of the United States
after the conclusion of the Mexican American War. Texas remains at its
full size extending into Wyoming. Acquisition of land to complete the
nation's westward thrust toward "Manifest Destiny" is shown by treaties
with foreign nations between the years 1783 and 1848. Free and slave
states are listed in square miles and acres. The map well illustrates the
rapid growth of the nation through the mid-point of the nineteenth
century as well as the concerns caused by the slave/free state issue.*

PLATE 40

Map of the United States and Mexico by Johnson & Browning.
Published by Johnson & Browning under the direction of Col. Carlos
Butterfield in *United States and Mexican Mail Steamship Line, and
Statistics of Mexico*; New York, 1859. 32" x 38".

*The large brightly-colored, detailed map highlights steamship routes be-
tween Europe, Africa, and America. The Gulf of Mexico ports primarily
used were New Orleans and Mobile. The table lists distances. The map
was issued just prior to Civil War, when steamships were beginning to
compete with railroads for western commercial and passenger traffic.*

well-armed abolitionist bands that were just as ardent in their
intentions. In this tinderbox, fighting was all but inevitable.
On May 21, 1856, a southern mob attacked the free-state
stronghold of Lawrence, Kansas, destroying property and intimi-
dating the population. In retaliation three days later, radical abolitionist John Brown led raiders against a pro-slavery settlement
at Pottawatomie Creek. There he and his followers butchered
five men with swords and axes before the eyes of their horrified
families. These events set off a running guerilla war between
pro-slavery men and free soil "Jay Hawks" that would eventually
claim more than two hundred lives. Gradually with sheer numbers, northerners gained the upper hand and many southerners
gave up in despair. John Whitfield sold off what property he
could and fled with his family to Texas. There he shared with
his new neighbors harrowing tales of abolitionist atrocities
(Plate 40).

The events in Kansas accelerated and intensified political
polarization. Even in the halls of government in Washington
physical violence began to replace discourse. In the wake of the
Pottawatomie massacre an inflammatory speech by Massachusetts Senator Charles Sumner provoked a brutal beating on the
senate floor by South Carolina Congressman Preston S. Brooks.
Many northerners insisted that they followed a higher law than
the Constitution and vowed to disregard the 1857 Dred Scott
decision that confirmed the South's interpretation of slavery.
Radical abolitionists like William Lloyd Garrison and states'
rights fire-breathers like William Lowndes Yancey heaped violent
abuse on one another in speeches and newspaper articles. In
1854 disparate political elements in the North established the
Republican Party, which steadily gained power as a result of the
continuing crisis. Southerners were appalled by the new party,
which they believed represented only northern interests.

Texas was not immune to the polarization and political
strife. Beginning around 1857 the extreme state rights wing of
the Democratic Party assumed control in Texas. Conservative
Unionists like Sam Houston and J.W. Throckmorton found
themselves increasingly out of step with the voters. Few Texans,
however, were ready to advocate extreme measures like secession—until events in Virginia began to change their minds.

In October 1859, John Brown, clandestinely supported by
prominent northerners, hatched a plan to incite the slaves of the
South to insurrection. He and his followers attacked the federal
arsenal at Harper's Ferry in order to secure a supply of arms. A
force of marines commanded by Colonel Robert E. Lee quickly
defeated the raid. Brown was convicted of treason and hanged.
He immediately became a martyr for the anti-slavery cause and
canonized as a hero throughout the North. The lionization of a
man whose avowed purpose was to massacre law-abiding citizens
and "bathe this land in blood" outraged southerners.

Texans now had reason to fear that the election of the
Republican Party to national office would not just entail a loss of
political power but would also threaten their lives and property.
The danger presented by Jay Hawk raiders or abolitionist fifth
columnists aimed at agitating slaves to bloody rebellion seemed
all too real. In the summer of 1860 mysterious fires broke out in
several North Texas towns. Citizens formed vigilance committees
and militia companies to guard against insurrection and arson.
Not surprisingly, fearful and over-alert citizens readily found
scapegoats. Before the panic subsided numerous whippings and
lynchings of slaves and suspicious strangers occurred.

Yet, the real cause of anxiety in Texas that summer was
the forthcoming election, which featured a three-way battle between a fractured Democrat Party, a cobbled-together Unionist
party, and an abolitionist-infested Republican Party that had

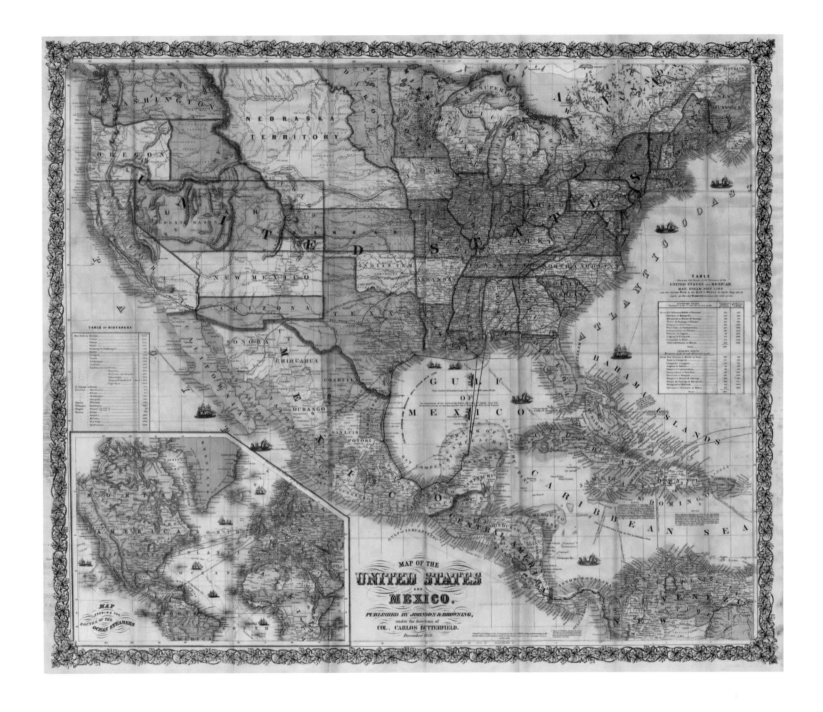

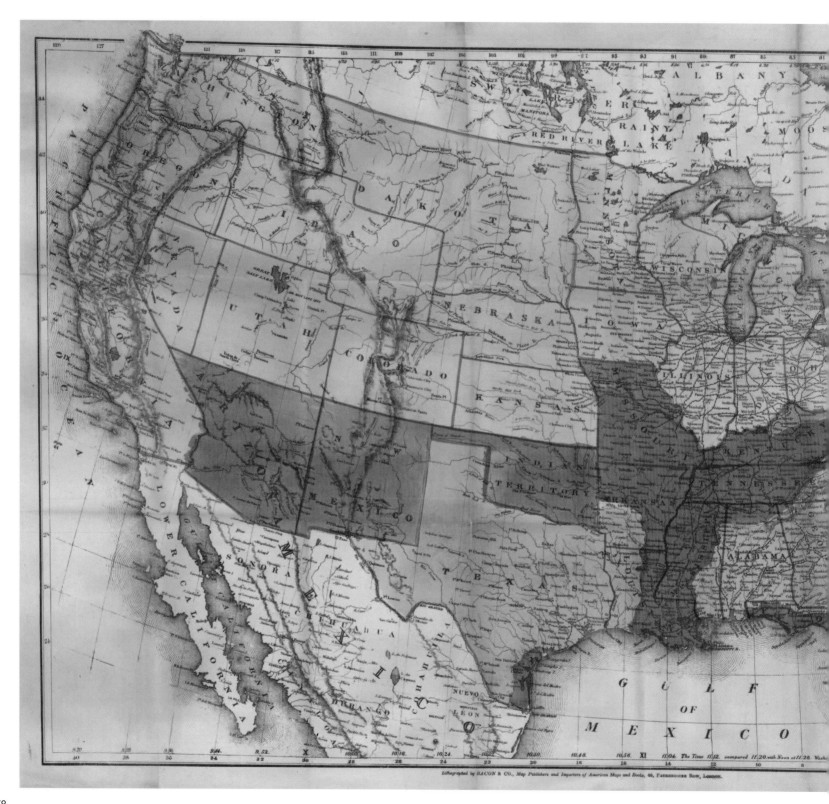

Lithographed by BACON & CO., Map Publishers and Importers of American Maps and Books, 48, PATERNOSTER ROW, LONDON.

EXPLANATION.
Distances on the Railways are
indicated thus ...52999......
Double Tracks ....
Titles & width ...... space a foot A ½ foot
of gauge
The width of gauge of all
Railways on this Map except
where otherwise noted is 4F84 in.

MAP OF THE
UNITED STATES,
Showing the Territory in Possession of the Federal Union.
January, 1864.

EXPLANATION.

	Area.	Population.
Territory claimed by the Confederates in 1861 (all the Slave States except Delaware)—PURPLE, GREEN, and YELLOW.	1,222,385	Free, 8,398,456 Slave, 3,949,620
Territory in the Military possession of the Confederates in 1861—GREEN and YELLOW	814,432	10,422,000
Territory Reclaimed from Rebellion by the Federal Union—GREEN.	302,000	4,718,000
Remaining in possession of the Rebels January, 1864—YELLOW.	512,422	Free, 5,265,000 Slave, 2,493,000

PLATE 41

Map of the United States Showing the Territory of the Federal Union, January 1864 by George Washington Bacon. Published by J. Snow in James William Massie's *America: The Origin of Her Present Conflict*: London, 1864. Lithography by G. W. Bacon & Co., Map Publishers and Importers of American Maps and Books . . . 18" x 27.5".

The large folding colored map, showing the areas under rebel control, reflects how isolated the Trans-Mississippi Department under Confederate General E. Kirby Smith had become in 1864. G. W. Bacon was the British publisher for J. H. Colton, in 1862. Bacon reused Colton's plates in his "Shilling Series of War Maps" (1862) and Colton's General Atlas (1863). Texas cotton, traded for military supplies, medicines, and consumer goods across the Rio Grande at Brownsville and Laredo was crucial to the Trans-Mississippi Department. Texans had more access to foreign imports than other federally blockaded states. The Civil War did not end in Texas until June 2, 1865.

nominated Abraham Lincoln. For Texans the issue of whether to allow slavery in new territories had evolved in their minds into a matter of simple survival for the southern way of life. With Republican victory, only two alternatives remained: would Texans attempt to defend themselves within the framework of the federal government or would state leaders exercise what the vast majority of southerners believed was the constitutional right to depart the union?

The answer was not long in coming. Lincoln's election in November sparked spontaneous demonstrations in favor of secession throughout the state. Prominent politicians and civic leaders took up the cause and called for a convention to meet in Austin on January 28, 1861, to discuss disunion. Among the few who opposed precipitate action was Governor Sam Houston, who almost alone realized that leaving the union would mean war. News that South Carolina had "gone out" on December 20, quickly followed by five sister states of the lower South, dashed any hope that Texans might avoid secession.

Delegates in Austin passed by a 166-to-8 vote an ordinance that repealed Texas' 1845 annexation and restored the state to independence. The only concession to Houston and the other dissenters was the requirement that voters ratify secession in a February 23 state-wide referendum. In the meantime the convention sent a delegation to Montgomery, Alabama, where the other seceding states had set up a provisional government for a new republic. Back in Texas the vote was 46,154 to 14,747 in favor of secession. Printers rushed to their presses to compose new maps emblazoned defiantly with the words—Confederate States of America (Plate 41).

SUGGESTED READINGS

Campbell, Randolph B. *An Empire for Slavery: The Peculiar Institution in Texas, 1821-1865*. Baton Rouge: Louisiana State University Press, 1989.

Olmstead, Frederick Law. *A Journey Through Texas: Or a Saddle-Trip on the Southwestern Frontier*. Austin: University of Texas Press, 1978.

Spurlin, Charles D. *Texas Volunteers in the Mexican War*. Austin: University of Texas Press, 1998.

Winkler, William, ed. *Journal of the Secession Convention 1861*. Austin: The Texas Library and Historical Commission, 1912.

POST-CIVIL WAR TEXAS HISTORY 1865 TO 1900

DAVID COFFEY

With all due respect to the Alamo and the other icons of Texas independence, the dynamic period that stretched from the close of the Civil War to the turn of the twentieth century contributed more to the Lone Star State's historical image and cultural identity than any other era. During this time Texas returned to the union and then moved slowly away from the Old South as it reaffirmed its uniqueness within the United States. This pivotal epoch gave rise to the cowboy and the cattle drive, buffalo hunters and railroads, unreconstructed Democrats and infamous outlaws. It witnessed the last bitter battles between the army and the Indians for control of the vast frontier. And it spawned a new brand of populist politics that brought sweeping change to America. In just thirty years, from 1870 to 1900, Texas population grew by more than 2.2 million—from 819,000 to surpass three million. From Reconstruction, through the Gilded Age, and into the Progressive Era, Texas redefined itself in a typically big way (Plate 62).

Few eras can be traced to an exact date of initiation, but post-Civil War Texas has Juneteenth—June 19, 1865—the day Union General Gordon Granger disembarked at Galveston to declare all acts of the Confederate government in the state null and void. General Granger then announced the emancipation of Texas slaves. Although it took years in some cases to liberate them all, black Texans have celebrated this date of deliverance each year since. The arrival of federal troops also ushered in the first phase of Reconstruction.

As in other former Confederate states, most whites in Texas resisted Reconstruction and took full advantage of President Andrew Johnson's lenient, sympathetic approach to reinstate the status quo antebellum. Ex-Confederates and planter elites held sway. Employing a combination of obstinacy, arrogance,

intimidation, and legislation, conservative elements attempted to maintain the old order despite the realities of emancipation. The federal army had too much on its hands to protect the newly freed, and the new Freedmen's Bureau lacked agents and the teeth to do the job that needed to be done. Both institutions suffered attacks and intimidation. To a vengeful United States Congress, this overtly unrepentant South quickly grew intolerable. Dominated by Radical Republicans, Congress imposed its own brand of Reconstruction, which divided the old Confederacy into five military districts to be administered by army generals, whose authority superseded that of elected officials. Texas and Louisiana formed the Fifth Military District, commanded initially by fiery General Philip H. Sheridan, a bona fide Union hero and, next to William T. Sherman, probably the most despised Yankee alive.

Military Reconstruction placed new demands and restrictions on an already-embittered people. Sheridan removed recently elected Governor James Throckmorton and replaced him with Unionist Elisha Pease. Congressional retribution included the disenfranchisement of some 10,000 Texas men with ties to the Confederate States of America. At the same time, almost 50,000 African Americans registered to vote, and most of these would exercise this new right of citizenship in coming elections. A constitutional convention dominated by Union Democrats and Republicans produced a document that protected the civil rights and voting rights of freedmen and promoted public education. The resulting elections, which excluded conservative powerbrokers, revealed a dramatic shift. Edmund J. Davis, a Republican and former Union general, won a narrow victory, with black voters accounting for some ninety percent of his total. Davis moved with alacrity to address rampant crime and lawlessness and to protect his black constituents by establishing a state police force and militia. Having produced an acceptable constitution and with a pro-Union governor in control, Texas was readmitted to the Union in March 1870. The following month General J. J. Reynolds, who replaced Sheridan, returned the state to civilian control (Plate 42).

As Washington's enthusiasm for Reconstruction waned, imposed restraints lapsed, and former Confederates regained the franchise. The old guard stormed back into power in Texas and across the South. Democrats soon reassumed the upper hand, and in the gubernatorial election of 1873 Richard Coke almost doubled Davis' total, promising the return of conservative rule. Republicans, despite overwhelming opposition, challenged the results on a technicality, but before the process could play out armed Democrats seized the capitol. When President U. S. Grant failed to intervene, Davis had no choice but to step aside (Plate 44).

Conservative "redeemers" systematically overturned most of the Republican agenda. The Democrats forged a new, highly restrictive constitution that embodied the principles of small government, limited executive powers, and white superiority. For black Texans, the promise of Reconstruction soon gave way to the reality that very little had changed for them.

While Texas Latinos and African Americans faced hard times, the great indigenous peoples—the Comanche and Kiowa—who lorded over the western plains for centuries, felt the full force of the Texas Rangers and the United States Army. The departure of federal troops from the Texas frontier during the secession crisis, coupled with the state's inability adequately to defend its frontier during the Civil War and Reconstruction, allowed aggressive Comanche and Kiowa warriors to prey on the fringes of Texas civilization, killing and burning, and taking human hostages and livestock in alarming numbers. Along the Mexican border, Lipan and Kickapoo raiding parties, taking advantage of the various distractions and an international boundary, drove stolen cattle into Mexico. In places the settlement line

PLATE 62

Official Texas Brags Map of North America by John Randolph. Published by John Randolph: Houston, 1948. Drafting by Mark Storm. 17" x 23".

After WWII, Texas pride was greater than ever. This illustration reflects that inflated view of self. This humorous map is a play on the Texas Centennial Map a decade earlier. The Centennial of Statehood celebration in 1946 was not as grand as the independence festivities. The Statehood Commission, formed in 1941, had a series of programs promoting economic and social improvement. A special U.S. stamp was issued and a national exhibition was held in the Library of Congress. This lighter side of statehood was probably a welcome relief to the more somber affairs.

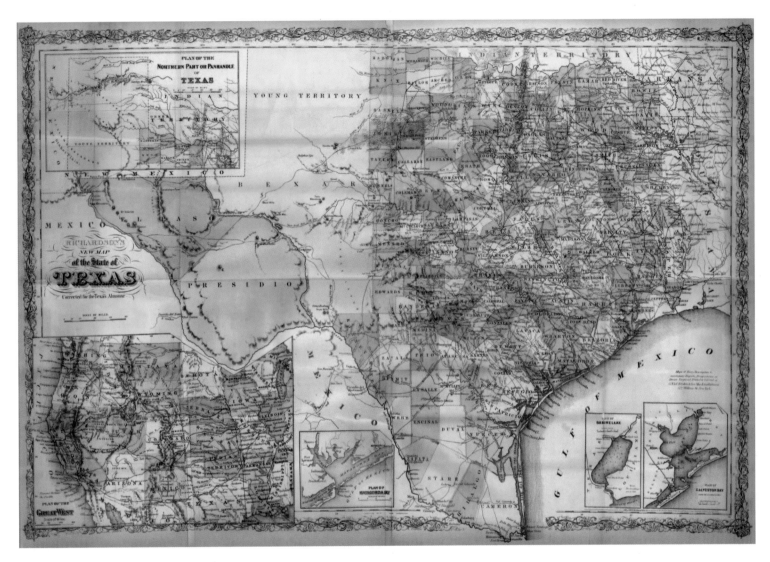

PLATE 42

Richardson's New Map of the State of Texas Corrected for the Texas Almanac by George W. and C. B. Colton & Co. Published by *The Galveston News* in *Texas Almanac for 1867*; Galveston, 1866. 18" x 26".

Col. Alfred H. Belo joined the Galveston News in 1866 upon his return from the Civil War. Richardson resumed printing the Texas Almanac in 1867. The post-war articles primarily concentrated on immigration to Texas, as well as business opportunities such as stock raising, cotton production and the growth of the railroads. Insets: Plans of Galveston Bay; Sabine Lake; Matagorda Bay; Great West and Panhandle. The western portion of the state was composed of vast undeveloped land districts: Young Territory, Pecos, Presidio, Béxar, and El Paso.

PLATE 44

Pocket Map of the State of Texas by Charles W. Pressler and A. B. Langermann. Published in Pocket Map; Austin, 1879. 26" x 27".

Pressler immigrated to Texas from Prussia in 1846 with the Adelsverein German settlement project. Jacob DeCordova hired Pressler in the Texas General Land Office to revise his highly successful 1848 Texas map. A treasury warrant paying Pressler a monthly salary as draftsman is shown in corner of frame. Pressler, like his supervisor, used actual land office surveys to update this edition. Pressler's map was first issued in 1858 as "Travelers Map of the State of Texas." Pressler's 1879 edition was a reduced pocket version. Railroads and related land grants are a predominate feature. Pressler had served as an officer in the Confederacy and after the war also worked on U. S. Government surveys. Pressler returned to the Land Office and retired in 1899.

PLATE 45

U. S. Telegraph Lines in California, Arizona, New Mexico and Texas in Charge of the Chief Signal Officer, U.S.A. by U. S. Government; Washington, D.C., 1879. 9" x 13".

The first Texas telegraph office opened in Marshall in 1854, with connections to New Orleans through Shreveport. The line soon was extended to Houston and Galveston. Austin was added in 1862. The Southwestern Telegraph Company, the first commercial company, was consolidated into the Western Union Telegraph Company, which began operations in Texas in 1866. By 1874 Western Union controlled ninety percent of the state's telegraph lines. The separate military telegraph system linked San Antonio with western outposts in 1876. The Galveston News first used the telegraph to wire news reports to its office in Dallas in 1885.

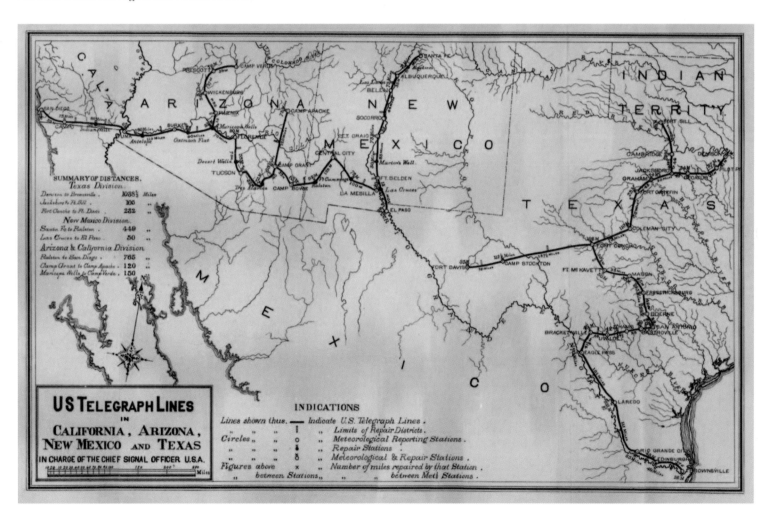

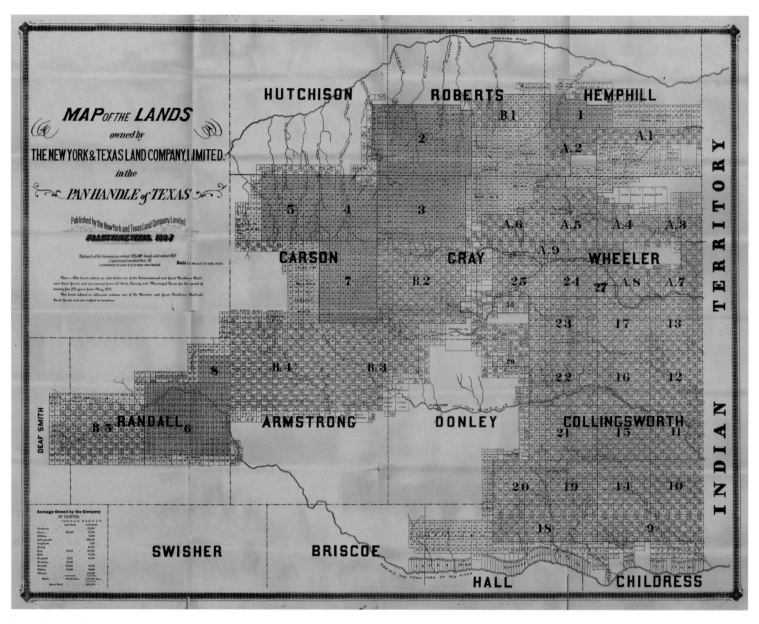

PLATE 47

Map of the Lands Owned by the New York & Texas Land Company, Limited by Gast, A.& Co.; St. Louis, 1882. Published by the New York and Texas Land Company Limited, Palestine, Texas. Lithography by A. Gast & Co., St. Louis. 28" x 36".

The New York and Texas Land Company was the transferee of the International-Great Northern Railroad Company. The three million-plus Panhandle acres were sold to ranchers and homesteaders for twenty years. Six hundred thousand acres became the Franklyn Land and Cattle Company (Carson, Grey, Hutchinson & Roberts counties), locally known as the "White Deer Lands." Another two-and-a-half million acres in West Texas was acquired from the state at cost of fifty cents per acre under the Land Act of July 1879. The company at one time held property in fifty-one Texas counties. Company crews surveyed and fenced much of the Panhandle ranch land. By 1900 most had been fenced with barbed wire and improved, allowing for sales to third party at price between $1.75 to $3 per acre. This map located those company tracts east of present-day Lubbock in Crosby, Dickson, Garza and Kent counties.

receded a hundred miles. Those who stayed did so at great peril, usually by working with neighbors to construct civilian forts for their mutual defense.

Reconstructed Texas was unique among the former Confederate states in this regard. Once federal authority returned, army officers faced not only the immediate demands of Reconstruction but also the need to subdue the Indian menace on the plains and stabilize a lawless and unruly border. Reduced from its 1865 high of more than a million men to a small professional force with an impossibly split mission, the army itself had to be redefined during the postwar years. By 1867 the army had reoccupied some of its abandoned posts in the TransPecos West (Forts Bliss, Davis, and Stockton) and in the southwest and along the Rio Grande (Forts McKavett, Clark, Duncan, and McIntosh), and commenced construction on a new line of forts along the immediate western frontier (Forts Concho, Griffin, and Richardson). Returns for the year 1867 showed more than 2,000 officers and men garrisoning fifteen posts on the Indian frontier and along the Mexican border. The returns also revealed the presence of newly constituted "Colored" regiments such as the Ninth Cavalry, which consisted entirely of African-American soldiers and white officers. These "Buffalo Soldiers" and their brethren in the Tenth Cavalry, and the Twenty-fourth and Twenty-fifth Infantry, served in Texas with great valor under some of the most hostile conditions soldiers of this period had to face (Plate 45).

Despite the return of federal troops, Comanche and Kiowa raiders remained a problem. President Grant's failed peace policy—a misguided attempted to "civilize" Indians under the guidance of well intentioned civilians—angered Texans and frustrated soldiers, who found their options for dealing with Indian activity too restrictive. The United States Army's Commanding General, William T. Sherman, although no supporter of Grant's peaceful approach, tended to downplay cries from aggrieved Texans as hyperbole, until during an inspection tour of Texas he almost became the victim of a bloody raid. In May 1871 Sherman and a small escort passed through Young County en route from Fort Griffin to Fort Richardson. Hours later, along the same trail Sherman had traveled, a band of warriors led by Kiowa chief Satanta attacked a wagon train, killing and mutilating most of the teamsters. The so-called "Salt Creek Massacre" destroyed the peace policy and opened the way for an all out military solution to the "Indian problem."

Sherman unleashed the army in what became the most active and brutally effective phase of the Indian Wars. Over the next seven years, beginning in Texas and spreading over the Great Plains, the army brought near total subjugation to the greatest of the North American tribes. Following the Salt Creek Massacre, the army, most notably, Colonel Ranald S. Mackenzie's Fourth Cavalry, campaigned relentlessly, striking deep into the Panhandle, where it was assumed no organized force could operate. Although Mackenzie failed to destroy the Comanche, he did put them on the defensive. With the plains tribes temporarily subdued, the army ordered Mackenzie and his Fourth Cavalry to Fort Clark, where cross-border banditry and Indian raids demanded aggressive action. In May 1873, with only a verbal implication from General Sheridan as his authority, Mackenzie led four hundred troopers from the Fourth Cavalry and two dozen Seminole-Negro scouts across the Rio Grande, striking Kickapoo and Lipan villages some forty miles into Mexico. Constantly on the move without sleep for sixty hours, Mackenzie's raiders posed an outright affront to Mexico. The raid could have gone terribly wrong. As it was, the soldiers destroyed their largely undefended targets, killing nineteen and capturing about forty women and children, and Mackenzie somehow avoided an international incident. As a result of the Fourth Cavalry's efforts, depredations by Kickapoo, Lipan, and Mescalero operating out of Mexico slackened significantly. Within two years most of the Kickapoo had accepted deals to relocate to reservations in Indian Territory. Mackenzie, a New Yorker by birth and a Yankee war hero, quickly became one of the greatest military figures in Texas history.

Now the army's frontier troubleshooters, Mackenzie and the Fourth Cavalry, returned to West Texas in preparation for a massive five-pronged campaign—the Red River War—to crush the Comanche, Kiowa, and Cheyenne bands that had risen defiantly, lashing out at ranchers, railroaders, and buffalo hunters across northwest Texas and Kansas. Without doubt, the Indians had major grievances. The government's failure to honor treaty commitments, unscrupulous contractors whose corruption prevented promised goods such as food and blankets from reaching the reservations, and the Indians' own cultural demands—which emphasized hunting and warrior prowess—all combined to ignite the uprising.

Another factor contributing to Indian unrest was the wanton and utterly demoralizing slaughter of the sacred buffalo, which provided essential food, clothing, shelter, and tools to the plains tribes. One expeditious way to eliminate the Indian threat, Washington quickly realized, was simply to eliminate the buffalo. The great herds already faced challenges stemming from railroad construction that cut the bison range, but government encouragement and the lucrative buffalo robe trade brought out professional

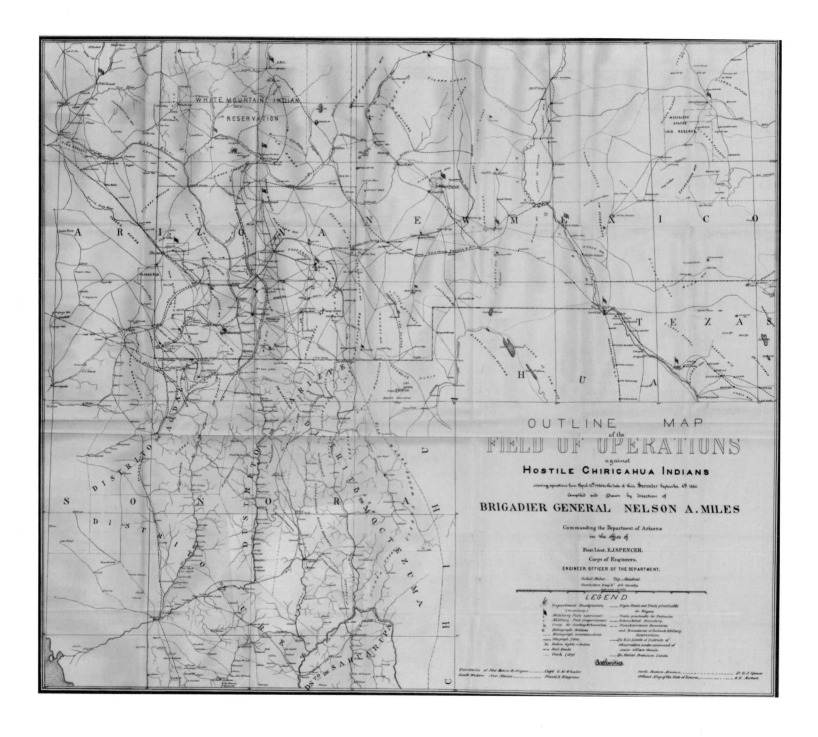

hunting teams and sportsmen in huge numbers, and the effect proved devastating. Buffalo that once numbered in the millions faced extinction by the early 1880s. The Indians retaliated, attacking hunters, work camps, and railroad gangs. In one particularly famous incident, a band of Comanche, Kiowa, and Cheyenne warriors, including the charismatic Quanah Parker, whose white mother was captured as a child and later famously rescued by Rangers, struck an old trading post and buffalo hunter camp on the Canadian River called Adobe Walls. But the well-armed and well-protected hunters employed their high-powered Sharps rifles with deadly effect, driving the raiders away with heavy losses (Plate 47).

During the fall and winter of 1874 five separate army columns converged on Indian strongholds hidden in the vastness of the Panhandle. The largest action occurred in late September when Mackenzie's column descended undetected into Palo Duro Canyon and, in an awesome display of total war, swept through a large village, burning hundreds of lodges and large quantities of stores, and capturing 1,500 ponies, more than a thousand of which Mackenzie ordered destroyed to keep them from being recovered. Throughout the winter army patrols kept up the pressure. Deprived of food, clothing, and shelter, one by one

PLATE 50

Outline Map of the Field of Operations Against Hostile Chiricahua . . . compiled . . . by Brigadier General Nelson A. Miles by E. J. Spencer, 1st Lt. U. S. Army. Published by U. S. Government; Washington, D.C., 1886. Lithography by American Graphic Company, New York. 29" x 25".

During the Civil War, Nelson Appleton Miles of Massachusetts rose to the rank of Major General of Volunteers, after displaying remarkable leadership skills on the battlefield. After winning the Medal of Honor he served as commander of Fort Monroe where Jefferson Davis was confined. Next, he was appointed assistant commissioner of The Freedmen's Bureau for North Carolina. He married the niece of Senator John Sherman and General William T. Sherman. After the war, he received a commission in the regular army as colonel and was posted to western campaigns. In 1874 he had a successful expedition against Indians in the Texas Panhandle during the Red River War. In 1877 he defeated Crazy Horse at Wolf Mountains. That same year he beat Chief Joseph's Nez Perce Indians at Bear Paw Mountains. He assumed command of the Department of Arizona in 1886 after a promotion to brigadier general. He accepted Geronimo's surrender at Skelton Canyon that September. He retired as a lieutenant general in 1903 after service during the Spanish American War (Puerto Rico), the Philippines Insurrection, and Boxer Rebellion. This Indian Campaign map of his southwestern service coincides with the end of the hostilities in Texas.

and group by group once-proud Indians moved onto reservations. By spring 1875 it was all over. Quanah Parker and the remnants of his band were among the last to surrender. The Red River War's total success opened new lands to farming, ranching, and railroads.

Isolated Indian activity and border unrest kept the army busy over the next few years, most notably in 1880 when Apaches under Victorio sparked serious concern in the Trans-Pecos but were forced out of the state to meet their demise in Mexico by the tenacity of the Buffalo Soldiers of Colonel Benjamin Grierson's Tenth Cavalry (Plate 50).

Enterprising cattlemen did not wait for Indian pacification. At the close of the Civil War millions of longhorns roamed Texas from the Rio Grande to the high plains. Since the state had escaped the destruction visited upon the rest of the South, it found itself uniquely positioned to supply beef to a rapidly growing and hungry nation. Texas cattle would be used to repopulate depleted herds and to open new ranches that followed in the wake of Indian removal. Rugged and adaptable longhorns provided a relatively cheap source of meat. Before railroads penetrated the Lone Star State, the cattle drive offered the only way to get cattle to far-flung grazing lands or to market. It could be a dangerous and financially risky business, but potential profits made it worthwhile. Soon Texas cowboys—white and black—and Tejano vaqueros pushed large herds from South Texas over soon-to-be-legendary trails, first to Sedalia, Missouri, then on the Chisholm and Great Western to railheads at Abilene and Dodge City, Kansas. The Goodnight-Loving Trail swung westward along the Pecos and up through New Mexico to new ranch lands in Colorado, Wyoming, and Montana. Other spurs led to Arizona and California (Plate 43).

The era of the great cattle drives proved relatively brief as population, barbed wire, and farming brought an end to the days of the open range. Soon massive ranching operations, such as the King Ranch, the massive XIT, and the foreign-owned Matador Land and Cattle Company, and the arrival of the railroad and refrigeration made the drive obsolete. Still, the cattle drive and the cowboy left an indelible mark on Texas history and American culture.

The cattle industry bred conflict as well. Combined with simmering hatreds spawned by secession, Civil War, and Reconstruction, a volatile border, and a lawless frontier, Texas became a particularly violent place during the postwar years. Vigilantism, as well as factional and racial violence, became commonplace during Reconstruction and throughout the 1870s. Much of the

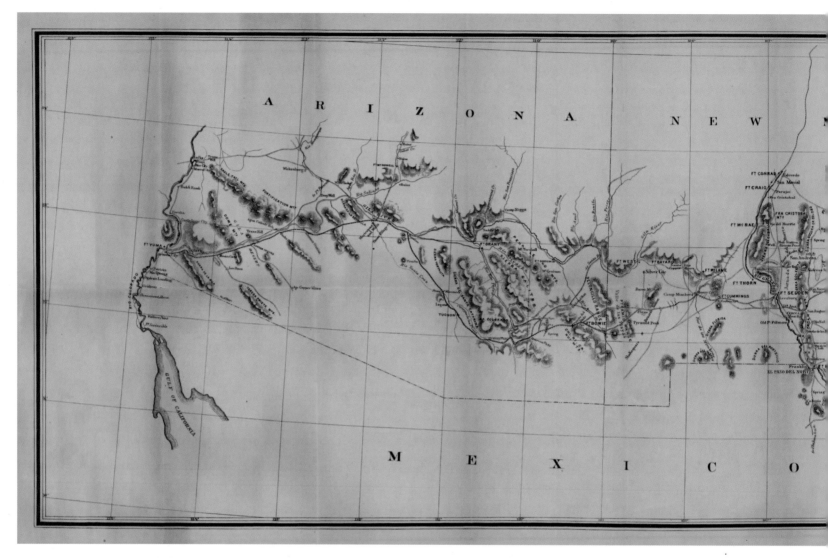

PLATE 43

Outline Map Showing a New Route from Texas to Fort Yuma, California, for Cattle Droves and Trains En Route to California by Charles W. Pressler, prepared under direction of Brevet Major General J. J. Reynolds, Commanding Department of Texas; 1870. Lithograph by Captain T. C. Overman.

The Texas cattle industry exploded between 1866 and 1890. At the end of the Civil War over three million head of unclaimed Texas longhorns were available for the taking. Valued at two dollars in Texas, the market was forty dollars a head in northern cities. In 1866, over 260,000 head *were driven to those markets. Contract drovers earned about a dollar a head for moving herds to market, normally numbering 3,000 head per trip. Abilene, Kansas, was the principal railhead up to 1873. The government map of a California route starting in Austin, Texas, to Fredericksburg, Ft. Mason, Ft. Concho, Ft. Stockton, Ft. Davis, Ft. Quitman, Ft. Bliss, to New Mexico, Ft. Selden, Ft. Cummings, Camp Mimbres and on to Arizona City (Yuma) via Ft. Bowie, Tucson, and the Pimas was to provide fresh beef to army posts. The route was never exploited beyond military suppliers.*

90

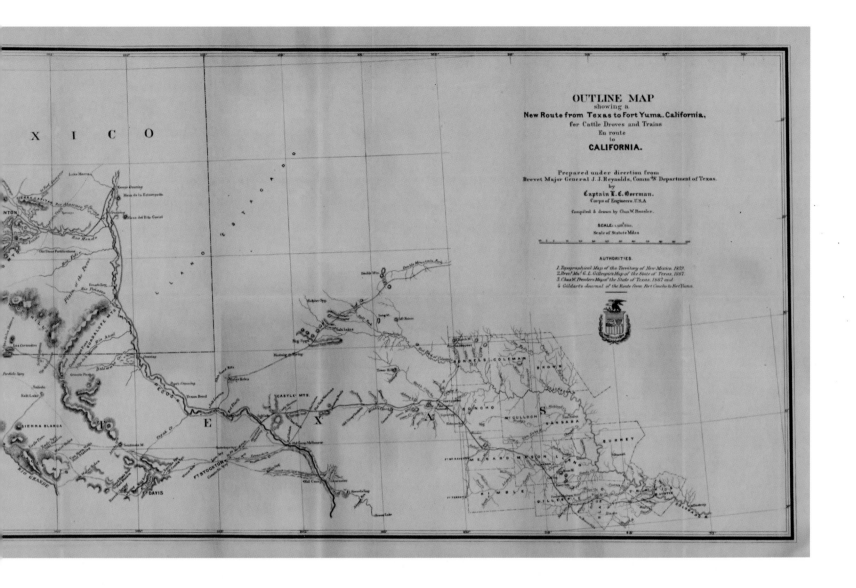

OUTLINE MAP
showing a
New Route from Texas to Fort Yuma, California,
for Cattle Droves and Trains
En route
to
CALIFORNIA.

Prepared under direction from
Brevet Major General J.J.Reynolds, Comm'g Department of Texas.
by
Captain T. C. Overman,
Corps of Engineers. U.S.A

Compiled & drawn by Chas W. Pressler.

SCALE: 1,500,000.
Scale of Statute Miles

AUTHORITIES.

1. Topographical Map of the Territory of New Mexico. 1859.
2. Brevt Maj'r G. L. Gillespie's Map of the State of Texas. 1867.
3. Chas W. Pressler's Map of the State of Texas. 1867 and
4. Gildart's Journal of the Route from Fort Concho to Fort Yuma.

PLATE 46

Map of the State of Texas Published by The International-Great Northern Railroad by Wm.P. Northrup & Co. Published by International & Great Northern Railroad in *Homes in Texas on the Line of the International-Great Northern R.R. 1880-1881 International Lone Star Route*; Palestine, Texas. 1880. Lithography: Wm. P. Northrup & Co., Map Engravers Buffalo, New York. 15.5" x 22"

This map was included as part of a railroad promotional guide issued by the I&GNRR. The railroad was part of the Texas "Missouri Pacific" line, serving east central Texas from Longview to Palestine. There it divided into two routes, one to Houston and the other to San Antonio. The promotional text advises anyone wanting to improve his position in life to go to Texas. Articles give detailed descriptions of 52 counties, with marketing company-owned land to emigrants. The draftsman effectively telescopes the railroad routes from the United States into a full scale map of Texas.

PLATE 48

Official Map of the State of Texas by A. W. Spaight, Commissioner
of Insurance, Statistics, and History. Published by Rand McNally & Co.
in Texas Dept. of Agriculture, Insurance, Statistics, and History, *The
Resources, Soil, and Climate of Texas*, December 1, 1882 (Galveston,
Texas: A. H. Belo & Co. Printers, 1882); Chicago, 1882. 332" x 34".

*A. W. Spaight, the Commissioner of Insurance, "Statistics and History"
during O. M. Roberts' 1882 governorship, prepared the detailed industrial map for public distribution to inform the world of the "exuberant
prosperity" the satisfied people of Texas were then enjoying. Taxable
property had increased from 280 million dollars in 1870 to 410 million
dollars in 1882. Tax rates had been reduced from fifty cents to thirty
cents per hundred dollars of property value. The only negative remark
was that the capital building had been destroyed by fire in 1881, and
the legislature was meeting in temporary quarters. Land grants were
issued for three million acres in the Panhandle for construction of another structure. The Capital Freehold Land and Investment Co., Ltd.,
incorporated in London in 1885, acquired most of that acreage to
form the XIT Ranch.*

violence fell on blacks and Tejanos. Wild cattle towns and frontier
outposts such as Fort Griffin tended to attract unsavory types,
and Texas certainly produced more than a few famous gun-
slingers such as John Wesley Hardin and King Fisher. The violence, lawlessness, and social upheaval of the postwar era made
the resurrection of the Texas Rangers as the state's chief law enforcement instrument inevitable. Rangers amassed a record of
success in quelling riots and feuds and other combustible situa-
tions, but they also gained a reputation for brutality and their
own disregard for the law. Still, Ranger heroes such as diminutive
Captain Leander H. McNelly became larger-than-life figures in
Texas history.

The military campaigns and the cattle boom set the stage
for the next, and perhaps greatest, agent of change—railroad expansion. Cattle could be driven to market if need be, but moving
goods and crops over the wide expanses of Texas proved dangerous and expensive. Like most of the southern region of the
United States, Texas lagged behind in railroad construction, but
the opening of the frontier and the desire to exploit economic opportunity finally induced frugal Texas politicians to support railroad construction with huge grants of public land. New
construction, though, posed problems for cattle and sheep producers who profited from open-range practices, while competition for new routes brought conflict and corruption. In 1873, the
Missouri, Kansas & Texas (Katy) Railroad, which entered Denison, Texas, linked with the Houston & Texas Central, connecting

the Great Plains to the Gulf of Mexico (Plate 46). The 1881
completion of the Southern Pacific connected a bourgeoning
Houston with distant El Paso, while later the Texas & Pacific
spawned a string of towns such as Abilene, Sweetwater, Big
Spring, Midland and Odessa. Railroad construction became a
major economic engine, providing jobs and markets for Texas
products such as timber—a growing industry in the postwar
years. Wherever the railroad went, population followed, and
commercial centers formed to support farmers and ranchers,
transforming Texas. By 1904 Texas boasted 10,000 miles of
track, more than any state in the nation (Plate 49).

Before oil became synonymous with Texas, agriculture,
namely cotton and cattle, remained the state's primary business.
Technological advances and the railroad allowed farmers to
spread onto arid western lands in the Cross Timbers and Panhandle regions. Although industrialization came slowly, timber production in East Texas, limited coal mining near company towns
such as Thurber, and the initial flirtations with petroleum exploration provided ample evidence that a more diversified economic
future loomed for Texans (Plate 48).

To realize the opportunities that lay ahead, Texas had to
address a major shortcoming: education. Support for public primary education ebbed and flowed during and after Reconstruction, and by the turn of the century was uneven at best. Although
early Texas leaders had provided for a state university in the
1830s, Texas Agricultural and Mechanical College at Bryan did
not open until 1876. It took another seven years before the new
University of Texas at Austin accepted its first students.

One of the motivating factors behind Texas A&M was a
growing interest in commercial farming and improved agricultural science. This impulse led directly to the first substantive
changes in Texas politics since Reconstruction. Conservative
Democrats tied to Old South ideals and dominated by former
Confederates such as John Reagan, Sam Bell Maxey, and
Lawrence S. "Sul" Ross, maintained a tight grasp on the state,
thwarting any real opposition to one-party rule. But a challenge
to the conservative stranglehold emerged in the 1880s as strug-
gling farmers began to demand more from their government. The
agrarian revolt traced its roots to the formation of the Patrons of
Husbandry, better known as the Grange, in 1867. Organized to
promote education and social interaction among farmers,
Grangers also began to emphasize economic cooperation. The
Grange gave way to the Texas Farmers Alliance, a more politi-
cized effort that reached out to other rising interests such as
prohibitionists and labor and crossed state lines to unite with

DENISON, TEXAS.
GRAYSON COUNTY.
1886.

PLATE 49

Denison, Texas, Grayson County, 1886 by Henry Wellge. Published by Norris, Wellge & Co.; Milwaukee, 1886. Lithograph by Beck & Pauli. 21" x 31".

Denison became a city after the Missouri, Kansas, & Texas Railroad crossed into Texas from Oklahoma in 1872. The company constructed a roundhouse, yard, and shops at the new town site on the Red River.

Named for George Denison, vice president of the railroad, the community rapidly grew. Henry Wellge's bird's-eye view of the city only fourteen years later, proudly reflects the double row of buildings along main street. The Houston & Texas Central gave service to Sherman and Dallas while the "Katy" went south to Fort Worth and east to Marshall. Noticeably absent is a courthouse on the square, since Denison was never the county seat for Grayson County.

similar groups. Without conceding racial equality the alliance also offered to incorporate black interests.

Democrats embraced elements of this progressive movement in an effort to harness any threat to one-party rule, and in Congress Texas Democrats pushed such populist issues as the expanded coinage of silver to promote inflation, tariff reform, and the regulation of interstate commerce. At the state level, Governor James Hogg moved away from the conservative pattern by championing antitrust reform and railroad regulation. What started as an attempt to organize and improve conditions for farmers—and eventually industrial workers—grew into a legitimate third party in the 1890s. The so-called Populists advanced a bold and revolutionary platform that included controversial ideas such as government ownership (or at least regulation) of transportation and utilities, a prohibition on national banks, an income tax, direct election of United States senators, the free coinage of silver, the Australian ballot, referendum and recall, and an eight-hour workday. With these measures, Populists sought to curtail the power and influence of giant trusts and the elites who controlled life in America. They also sought a better, truer form of democracy. Texas Democrats embraced many of these goals, at least in part, but most refused to support the controversial subtreasury proposal which, to their minds, threatened too much government control. Although the Populist Party collapsed, Democrats and Republicans alike adopted its principles, many of which would become reality in the coming decades.

As Texas closed the curtain on the nineteenth century, it retained much of its frontier identity and many of its ties to the Old South. But the state had evolved in dramatic ways. In 1900 a devastating hurricane all but destroyed Galveston, once the commercial capital of Texas. Yet Houston had emerged as a major inland port and railroad terminus uniquely positioned to reap the rewards of the coming petroleum boom. San Antonio, the state's largest city in 1900, remained as it had been for decades—a cultural and military center. Dallas was already establishing its profile as a market hub while Fort Worth capitalized on its "cowtown" legacy to attract huge meatpacking operations and other service industries. Railroads now linked the major population centers. The population had by pattern and politics become largely segregated and dominated by the white majority, but African-American and Latino communities continued to define themselves and contribute meaningfully to the state's development. Distinctive ethnic communities, while becoming increasingly assimilated, clung to their German, Czech, and Polish roots. And despite substantial growth in the cities, Texas remained, in 1900, a predominately rural and agrarian state, poised to play a starring role in the twentieth century.

SUGGESTED READINGS

Barr, Alwyn. *Reconstruction to Reform: Texas Politics, 1876-1906.* New Edition. Dallas: Southern Methodist University Press, 2000.

Haley, James L. *The Buffalo War: The History of the Red River Indian Uprising of 1874.* Abilene: State House Press, 1998.

Moneyhan, Carl H. *Republicanism in Reconstruction Texas.* College Station: Texas A&M University Press, 1980.

Utley, Robert M. *Lone Star Justice: The First Century of the Texas Rangers.* New York: Oxford University Press, 2002.

_____. *Frontier Regular: The United States Army and the Indian, 1866-1891.* Lincoln: University of Nebraska Press, 1973.

Disorder on the Border
1910-1923

Matt Walter

—◆◆◆—

Following the signing of the Treaty of Guadalupe Hidalgo in 1848, relations between the United States and Mexico remained generally peaceful for the next six decades. Except for a few territorial squabbles caused by the shifting of the Rio Grande River itself, the two nations cooperated in the mapping of the border and the division of water resources between the two countries. Otherwise, internal affairs remained the focus of both countries. The United States fought the Civil War and then concentrated on Reconstruction and westward expansion. At the same time, Mexico fought a war with France to defeat the puppet emperor Maximilian. Yet having won a war for independence, Mexico replaced Maximilian with the dictator Porfirio Díaz, one of the heroes of the war against the French. Díaz remained dictator of Mexico from 1876 until 1911 (Plate 54).

The Mexican Revolution of 1910 and World War I marked major turning points for relations between Mexico and the United States. The peace of the previous decades was replaced with disorder, incursions, and a general increase in border troubles and tensions. Fighting in Chihuahua between federal forces and revolutionaries led by Francisco "Pancho" Villa began in 1911. Both Mexican refugees and Americans living on the Mexican side of the border sought refuge in the United States, with many fleeing to the railhead at Marfa, Texas. Outlaws and bandits on both sides of the border took advantage of the chaotic situation to raid stores, steal cattle, and settle blood feuds. On the United States side a handful of sheriffs, Texas Rangers, and customs inspectors did their best to maintain order and protect the towns and ranches in the Big Bend area. They were assisted by the few federal troops assigned to the region. Less than fifty United States soldiers garrisoned Marathon, and a soldier or two guarded each post office. In February 1911, responding to repeated requests from local officials and Texas' governor, the federal government ordered one hundred troopers from the Third United States Cavalry to Presidio, Texas, to patrol more than five hundred miles of border. In January 1913, United States Customs Inspectors Joe Sitters and J.S. Howard, along with cattle inspector J. W. Harwick, arrested a Mexican national on the United States side of the border for smuggling and horse stealing. During the arrest, they were ambushed by the infamous bandit Chico Cano and his gang. Harwick died in the gunfight that ensued.

Despite this incident the United States Army withdrew most of the troops along the border later that month, leaving only twenty-five men from the Third Cavalry to patrol the Big Bend. In February 1913, Chico Cano and his men raided two more ranches on the United States side of the border, taking guns, saddles, ammunition, and horses. Brewster County Sheriff J. Allen Walton organized a posse, but Cano escaped. Texas Governor Oscar Colquitt wired President William Howard Taft requesting that more troops be assigned to protect the Big Bend, but the War Department told the president that the troops were needed more in other areas of the border.

The raids continued, and in 1915 the new Texas governor, James Ferguson, asked President Woodrow Wilson to station more troops in the region. Despite the governor's entreaties, the commander of the Southern Department, General Frederick Funston, argued that he had no troops to spare and that the problems in the Big Bend were basically civil matters that local law enforcement officers should handle (Plate 55).

One of the most significant battles of the Mexican Revolution took place in the Big Bend. Pancho Villa defeated Mexican federal forces led by General Salvador Mercado, forcing them to evacuate Chihuahua City on the 13th of December, 1913.

PLATE 54

**Situación de la Línea Divisoria entre Méxio y Los Estados Unidos . . .
Comisión Internacional de Límites, General Anson Mills** by Thomas
F. Perry. Published by Norris Peters Co.; Washington, D.C., 1911.
16" x 24".

*Anson Mills, after failing out of West Point, settled in El Paso. In 1858
he was a surveyor for the forts of Quitman, Davis, Stockton, and Bliss.
He platted the new community of Franklin in 1859, which eventually
became El Paso. After the outbreak of the Civil War, he accepted a
commission in the Union Army, where he remained until 1894, retir-
ing at the rank of brigadier general. Afterwards, he was appointed to
the Mexican/United States Boundary Commission, where he remained
until 1914. Under his direction, the border was reestablished on the
island of San Elizario, below El Paso. He pushed an international
dam along the Rio Grande River which eventually became Elephant
Butte Dam in New Mexico. He was instrumental in the Chamizal
Tract boundary negotiations with Mexico.*

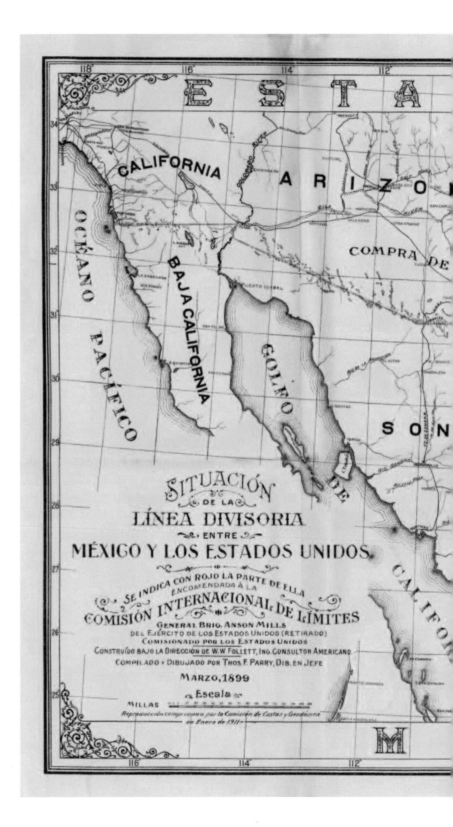

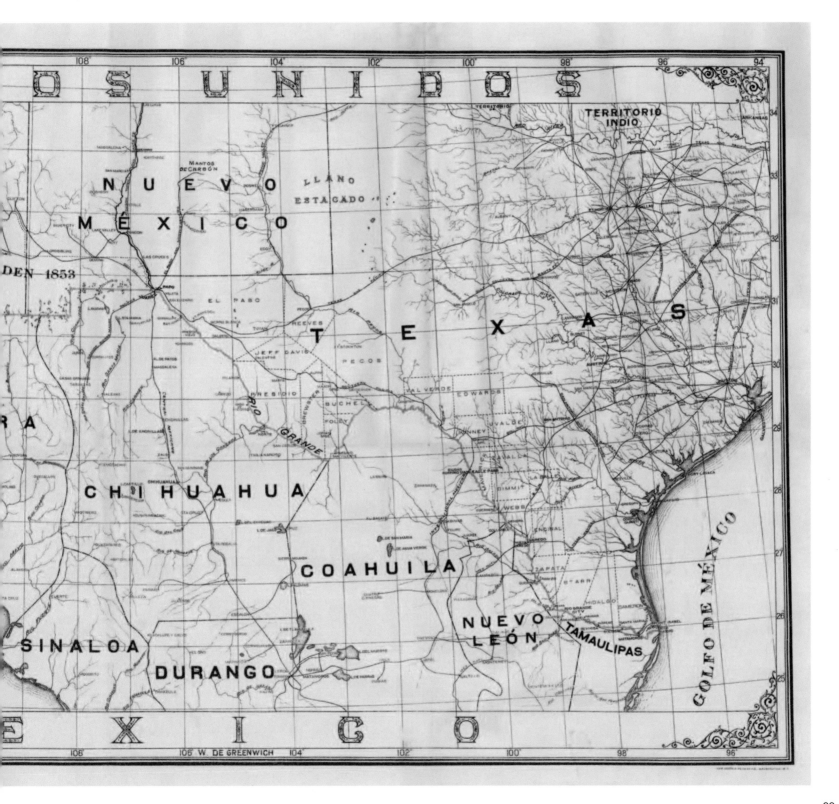

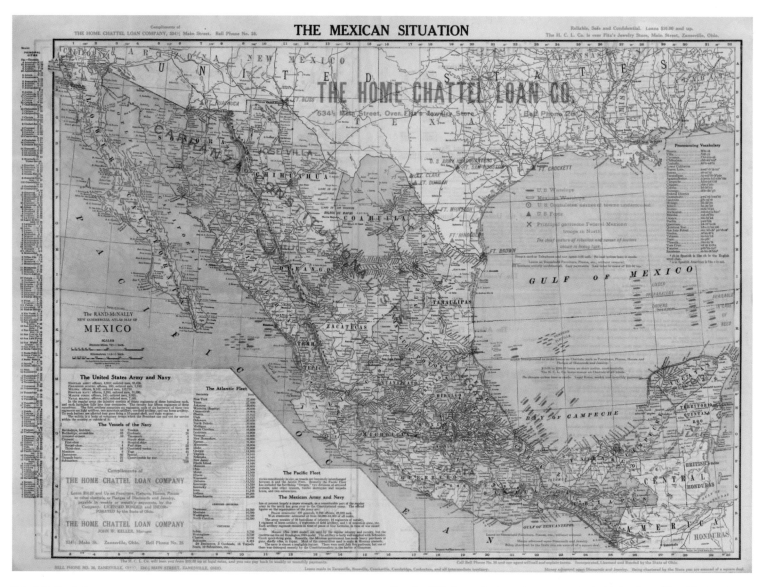

PLATE 55

The Mexican Situation by Rand McNally. Published by Home Chattel Loan Company; Chicago, 1914. 20.5" x 26".

The Mexican War situation map was created by the Home Chattel Loan Company by simply overprinting a standard Rand McNally map of Mexico. The edition adds warships, forts, Mexican garrisons, and military routes. The Mexican Revolution that began in November 1910 affected border Texans up to 1920. After a raid by men from General

Pancho Villa's revolutionaries was made in Columbus, New Mexico, in March 1916, the U.S .Government ordered General John J. Pershing to form a punitive expeditionary force. In May 1916, Mexicans crossed into Texas at Boquillas and Glenn Spring in the Big Bend. The U. S. National Guard was then called into Federal service. Military actions involving mechanized vehicles and airplanes paved the way for service in Europe a couple of years later.

Villa, leading his Army of the North, consolidated his hold on Chihuahua City and assumed the office of governor of Chihuahua. Villa also signed a contract with the Mutual Film Company of Los Angeles to have them film his battles, and a crew led by photographer Charles Rosher was sent to record the events unfolding in Chihuahua. After settling into the governor's palace in Chihuahua City, Villa sent two of his generals, Ortega and Natera, along with some 3,000 troops, to attack Mercado and his 5,000 federal forces in Ojinaga, just across the Rio Grande from the American town of Presidio. The United States, anticipating an increase in the number of refugees crossing the border, sent detachments of the Eleventh, Fourteenth, and Fifteenth Cavalry Regiments to patrol the border.

On December 30, 1913, Villa's forces, along with Rosher and his film crew, arrived at the outskirts of Ojinaga. On January 1, 1914, Villa's Army of the North engaged the Mexican federal forces protecting the city. Over the next four days the fighting did not go well for the *Villistas*, and Mercado's forces actually captured Rosher and his film crew. Villa, learning of the situation in Ojinaga, resigned as the governor of Chihuahua and assumed direct command of the rebel forces. During the evening of December 10 and the early morning hours of December 11, 1914, Villa used the cover of darkness to attack and rout Mercado's *federales*, who abandoned Ojinaga and fled into the countryside. The following day, Rosher and his liberated film crew filmed re-enactments of the Battle of Ojinaga, staged according to the *Villistas* version of what occurred in the darkness.

Pancho Villa's victory at the Battle of Ojinaga consolidated his control over a major portion of northern Mexico and made him one of the leading contenders for the office of president of Mexico. Villa became famous on both sides of the border and even met with General John J. "Black Jack" Pershing, the commander of the United States forces stationed along the border. American President Woodrow Wilson viewed Villa in a positive light, especially since Villa was the only revolutionary leader in Mexico who had not condemned the American takeover of the city of Veracruz in 1914. In April 1915, when Villa lost the Battle of Celaya to General Alvaro Obregón, leading the forces of presidential contender Venustiano Carranza, President Wilson, in a surprise move, withdrew his earlier support for Villa and endorsed Carranza, officially recognizing him as President of Mexico in October 1915. In November Wilson even allowed Carranza's forces to travel through the United States by train in order to attack the *Villistas* at Agua Prieta, across the border

from Douglas, Arizona. Feeling betrayed, Villa attacked Columbus, New Mexico, on March 9, 1916. In response, President Wilson ordered a "Punitive Expedition" under General Pershing into Mexico to defeat Pancho Villa.

The Big Bend region itself remained unprotected, and bandits continued to raid ranches and haciendas on both sides of the border. On May 5, 1916, Rodriguez Ramirez and Natividad Alvarez, known associates of Chico Cano, along with between sixty and seventy raiders (mostly from Mexico but also including several Texas outlaws) raided Glenn Spring and Boquillas on the United States side of the border. Glenn Spring was the home of C. D. Wood and W. K. Ellis, who employed around fifty local workers in their candelaria wax factory. There was also a small post office and general store there, and the community was protected by Sergeant Charles Smyth and six soldiers from the Fourteenth Cavalry. Boquillas was a smaller community situated directly across the river from the Mexican village of the same name. Jesse Deemer and his clerk, former army scout and Black Seminole Monroe Payne, ran a small store there. The raiders split into two groups, attacking both places around midnight. At Glenn Spring, the soldiers barricaded themselves inside an adobe hut, but they were forced out after a torch was tossed onto the thatched roof. Three soldiers were killed, but Sergeant Smyth and three others managed to escape. The outlaws also killed seven-year-old Tommy Compton while his father was hiding Tommy's younger sister with the family of a local worker. C. D. Wood heard the shooting from his house about two miles away. At first he thought it might be people celebrating *Cinco de Mayo*. When he realized that it was not, he and his neighbor, Oscar de Montel, grabbed their rifles and headed towards the fighting. They arrived too late to help, finding that the bandits had left with nine cavalry horses and loot from the store. In Boquillas, Deemer and Payne realized they were greatly outnumbered and did not put up any resistance. The outlaws looted their store. Around 10 A.M. on the sixth, the two outlaw bands regrouped and crossed back into Mexico, taking with them a great amount of loot and Deemer and Payne as hostages.

Coming so soon after Villa's attack on Columbus, New Mexico, the raids on Glenn Spring and Boquillas led to the "Second Punitive Expedition." Portions of the Eighth Cavalry, under the command of Colonel Frederick W. Sibley, pursued the outlaws, rescuing Deemer and Payne and recovering the stolen property.

Troops A and B of the Eighth Cavalry, under the command of Major George Langhorne, left El Paso and proceeded to

MAP OF THE UNITED STATES SHOWING

NATIONAL GUARD MOBILIZATION TRAINING CAMPS: NATIONAL ARMY CANTONMENT CAMPS; RESERVE OFFIC

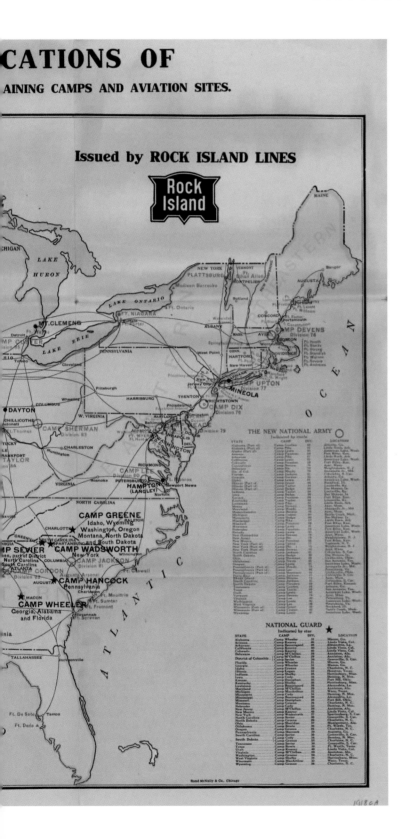

CATIONS OF

AINING CAMPS AND AVIATION SITES.

Issued by ROCK ISLAND LINES

Rock Island

PLATE 56

Map of the United States Showing Locations of National Guard Mobilization Training Camps; National Army Cantonments Camps; Reserve Officers Training Camps and Aviation Sites by Rand McNally & Co. Published by Rock Island Lines; Chicago. 1918. 22.5" x 37".

The European conflict that became World War I began in August 1914. The Germans sank the Lusitania in May 1915 bringing the United States closer to war. After General "Black Jack" Pershing began his punitive expedition into Mexico in March 1916, the country was in a state of war preparedness. After Germany sent the so-called coded Zimmermann Letter to their ambassador in Washington for the president of Mexico, Texans became outraged. The German letter promised assistance if Mexico would join the German/Japanese alliance. It was suggested that Mexico might regain territories lost after the Mexican American War. President Wilson declared war against Germany on April 2, 1917. Almost 200,000 Texans served in WWI. Over half the war deaths were due to the 1918 flu epidemic. Military bases and training camps sprang up all over the state, many of which remain today. The war ended November 11, 1918.

Marathon, Texas, then to Boquillas. Accompanied by reporters, journalists, photographers, and a film crew, all of whom rode in two Ford sedans, Langhorne and his mounted troops crossed over into Mexico on May 11, 1916. The two Fords and Langhorne's own chauffer-driven Cadillac touring car also transported grain for the horses.

The outlaws had divided their force and left Boquillas, Mexico, three days earlier. As he headed south, Langhorne received word that Deemer and Payne were being held at the village of El Pino, around one hundred miles south of Boquillas. Langhorne took the three cars and rushed ahead of the mounted troops, hoping to trap the kidnappers at El Pino. The bandits learned of the approaching soldiers and departed hurriedly the night before, leaving Deemer and Payne in the village. Langhorne's "mechanized cavalry" liberated the two hostages and recovered many of the items taken from Glenn Spring and Boquillas.

Langhorne spent a few more days in El Pino, hoping to capture some of the bandits and recover the stolen horses. A scouting party surprised a group of bandits who fled into the night, abandoning a wagon, seventeen horses and mules, nine rifles, and several saddles and packs. Upon learning of Langhorne's success, Colonel Sibley, who had also crossed the Rio Grande into Coahuila, determined that they had completed their assigned mission. On May 18, 1916, General Funston ordered the Second Punitive Expedition to return to the United States.

Three days later, Langhorne crossed back to the north side of the Rio Grande. The major's expedition had traveled more than five hundred miles during his eight days in Mexico. Without suffering a single casualty they had rescued Deemer and Payne and recovered most of the property stolen during the Glenn Springs and Boquillas raids. Though not on the scale of the more famous expedition led by Pershing, the Second Punitive Expedition proved, in many ways, far more successful.

By July 1916 more than 100,000 American troops, mostly from National Guard units, were stationed along the United States-Mexico border. Some of the border troops, such as the recruits who made up most of the army's Thirty-fourth Infantry division, came from as from as far away as Iowa and Minnesota. Other units, such as the Fourth Texas Infantry, were more familiar with the region. Among those Texans was Jodie P. Harris III, a member of Company I from Mineral Wells, Texas, who sent a series of hand-drawn postcards home to his family. Harris' postcards are full of humor and biting social commentary. Interestingly, he was also one of the first people to call for the Big Bend region to be designated as a national park (Plate 56).

In 1918, Congress recognized the troops assigned to the United States-Mexico border with a special Mexican Border Service Medal. With all the troops stationed on the border, the number of border raids and incursions decreased but did not end. Private Carl Albert, for example, was killed while chasing Chico Cano and his gang in March 1918. In June 1919, the Army Air Service began flying Dehavilland DH-4s from Marfa, Texas. In the following month, pilots Lieutenant Peterson and Lieutenant Davis crash-landed near Coyame, Chihuahua, Mexico, where they were taken hostage by Jesus Renteria and held for ransom. Local ranchers, attending the Bloys Cowboy Camp Meeting outside of Fort Davis, raised the $15,000 that Renteria demanded in ransom. Captain Leonard Matlack proceeded south, entered Mexico, paid Renteria half the ransom and then, when the hostages were handed over, refused to pay the second half of the ransom. Matlack returned safely with both of the pilots, and the secretary of war authorized Matlack to reenter Mexico to punish Renteria. On August 20, DH-4 pilots killed Renteria and three of his band from the air. With the mission accomplished, Colonel Langhorne, now commanding all United States forces in the Big Bend region, ordered Matlack to withdraw his detachment from Mexico. The following year, as General Alvaro Obregón became president of Mexico and began sending forces to police the Mexican side of the United States-Mexico border, the War Department issued General Order #15, reducing United States forces in the area. The Army Air Service left the Big Bend in 1921, and the border troubles finally ended when Pancho Villa was assassinated in 1923.

SUGGESTED READINGS:

Davenport, B. T. *Soldiering at Marfa, Texas: 1911-1945.* Kearny, Nebraska: Morris Publishing, 1997.

Harris, Charles Houston. *The Border and the Revolution.* Las Cruces: New Mexico State University Press, 1988.

Justice, Glenn. *Revolution on the Rio Grande: Mexican Raids and Army Pursuits, 1916-1919.* El Paso: Texas Western Press, 1992.

Katz, Friedrich. *The Life and Times of Pancho Villa.* Stanford: Stanford University Press, 1998.

Martinez, Oscar J. *Troublesome Border.* Tucson: University of Arizona Press, 1988.

Metz, Leon. *Border: The United States-Mexico Line.* El Paso: Mangan Books, 1989.

Perkins, Clifford Alan. *Border Patrol: With the United States Immigration Service on the Mexican Boundary, 1910-1954.* El Paso: Texas Western Press, 1978.

Romo, David Dorado. *Ringside Seat to a Revolution: An Underground Cultural History of El Paso and Juarez, 1893-1923.* El Paso: Cinco Puntos Press, 2005.

Welsome, Eileen. *The General and the Jaguar: Pershing's Hunt for Pancho Villa.* New York: Little, Brown and Co., 2006.

TEXAS IN THE TWENTIETH CENTURY

R E B E C C A   S H A R P L E S S

The story of Texas in the twentieth century is one of growth and urbanization. In 1900, the state's population was just over three million people, eighty-three percent of whom lived and worked on farms. One hundred years later, the population was almost 21 million, and fewer than three percent worked in agriculture. Texas was the first state in the South to have a majority of its population living in cities. These demographic changes reflect major shifts in population, economic factors, and culture.

In 1900, most Texans were either descendants of Anglo people from the American Southeast or African American. Groups of European immigrants and their descendants dotted the state, most notably in the so-called "German Belt" that stretched from the coastal plain to the Hill Country, and the 250 Czech communities across the eastern portion of Texas. Two major changes occurred in the years after 1910. World War I brought an end to European immigration and significant prejudice against German Texans, many of whom hastened their assimilation process to avoid attacks by angry Anglo Americans. And the Mexican Revolution forced thousands of immigrants north of the Rio Grande, spreading from San Antonio into Central Texas to escape the fear and violence in their native country. The cultural landscape of Texas in the early twentieth century, therefore, was not simply black and white but featured increasing diversity, particularly with the growth of the Mexican population.

At the dawn of the twentieth century, the average Texan spent a lot of time thinking about the weather, wondering when it would rain and whether the rain would be too little or too much, and following mules up and down long rows of cotton. Cotton dominated the state's economy. Throughout the eastern third of Texas, large and small farmers alike lived according to the vagaries of the world cotton market. For some, especially town dwellers, cotton brought prosperity. Small towns boasted fine mansions belonging to brokers and merchants, the middlemen who served as the link between the farmers and the world markets. For others, the white staple brought only poverty and sorrow, lives filled with hard work and little monetary reward. To the west, cattle ranches yielded supplies of beef but little to sustain people, and the counties of West Texas had small human populations.

Lumber was the biggest organized business in Texas at the beginning of the twentieth century. In 1901, John Henry Kirby consolidated fourteen sawmills into the Kirby Lumber Company, at that time the largest corporation in Texas. Between 1900 and 1910, the Texas lumber industry cut an average of two billion board feet per year. The 2.25 billion cut in 1907 set a state record that still stands. While the lumber industry decimated the pine forests of East Texas, the business models worked efficiently as long as there were trees, and lumber provided jobs for many Texans through the end of the 1920s.

In 1900, Galveston was among the most prosperous cities in Texas, receiving immigrants and shipping cotton out into the world market through its port. The city's prominence, however, would enjoy only a brief life in the new century. As Galveston had built its wealth from the sea, so it lost its wealth to the sea, as the storm of September 1900 slammed into the city and killed an estimated 6,000 people. The state turned its attention inland, and Galveston would never recover its previous glory. Another event took place on the Gulf Coast in 1901 that would instead set the tone for the new century: the massive oil strike at Spindletop, near Beaumont (Plate 53).

While nineteenth-century Texans were generally aware of oil deposits underground, they were at first hard-pressed to come

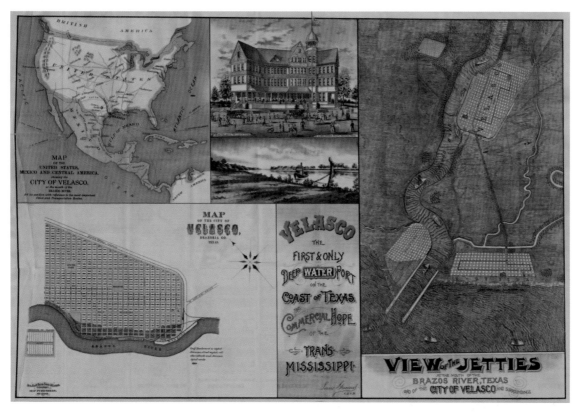

PLATE 53

Map of the United States, Mexico and Central America, showing the City of Velasco . . . View of the Jetties at the Mouth of the Brazos River . . . and Surroundings by Gast, August, Bank Note & Lithographing Co. Published in *Velasco the First only Deep Water Port on the Coast of Texas, Commercial Hope of the Trans-Mississippi*; St. Louis, 1892. 24" x 26".

The original city of Velasco in Brazoria County was founded in 1831. In 1875 the town was destroyed by a hurricane. The new town was established in 1891. A million-dollar real estate promotion accompanied the announcement of the resort hotel, Velasco. New rail service to Houston, steamer terminals, commercial warehouses and a deep water port followed. After the town grew, the government built a new lighthouse and granted a national bank charter. When another hurricane in 1900 destroyed both Galveston and the new city of Velasco, the town had grown to a population of 3000. The precinct was absorbed into Freeport's Brazosport industrial complex. This promotional brochure foretold a promising future as a recreational resort. The August Gast Bank Note & Lithographing Company published many maps for the Texas General Land Office.

up with ways to use the stuff. The people of Corsicana, site of the first major discovery in 1894, sprayed oil on their roads to keep the dust down. But the development of the oil- or gasoline-burning internal combustion engine changed everything. Vehicles that could move on their own—so-called "automobiles"—were becoming the rage at the turn of the century. Joseph Cullinan established the state's first commercial refinery to utilize efficiently the wealth of the Corsicana field, and the Cotton Belt Railroad became the first train line to run on oil rather than coal. In 1901, developers tapped the Spindletop dome and hit a gusher of unbelievable force. Speculators rushed in, and by the end of the year, the State of Texas had chartered almost five hundred oil companies. Large corporations such as the Texas Company (to become Texaco) and Gulf Oil built refineries in southeast Texas, and the new Houston Ship Channel rapidly developed into one of the busiest ports in the United States. With increased exploration, oil discoveries sprouted across much of the state, from the southeast to the Panhandle. By 1930, the Texas oil industry produced almost 300 million barrels a year, employed more than 13,000 people, and added more than $400 million to the Texas economy (Plate 57).

The development of the automobile had other ramifications for Texans, beginning early in the twentieth century. Henry Ford marketed the first Model T in 1908, and by 1916, when Texas began registering cars, there were almost 200,000 in the state. Texans began lobbying for better roads to run their new machines on, founding the Texas Good Roads Association in 1916. At first local wrangling prevented a unified system of roads from developing statewide, but by 1929, the state had more than 9,000 miles of hard-surfaced highways. Good roads enabled farmers to get their products to town more efficiently and also allowed them to work in town and live in the countryside.

Urban Texans, too, could move from place to place with greater facility. Texas society became more mobile and gradually less localized, as people could go to a variety of places for entertainment and shopping (Plate 58).

Even as Texans took to their cars, the Texas railroad industry continued development. Between 1900 and 1932, track mileage throughout the state almost doubled, with vast sections of South and West Texas receiving service for the first time. As the industry grew, it also consolidated. Although a number of independent lines continued, by the Great Depression, the Southern Pacific, Missouri Pacific, and Santa Fe owned or controlled more than seventy percent of the railroad mileage in Texas. They continued moving people and goods at ever-increasing speeds across the state.

By 1920, the agricultural system which had formed after the Civil War was beginning to change. Texas was at the forefront of agricultural reform efforts which taught farmers to care for their land and to diversify their crops. Furthermore, the development of irrigated agriculture in West Texas spurred the creation of large cotton farms, highly mechanized and capital intensive. Wheat and citrus made inroads in the Panhandle and Rio Grande Valley, respectively, but cotton cultivation took more acreage than all other commercial crops combined. Hard times loomed, however, as agricultural prices began to slide after World War I. Rural Texans joked that they were already so used to being poor that they did not notice the Great Depression.

Joking aside, few Texans were affected directly by the crash of the stock market in 1929, but the collapse of the nation's economy eventually made its way south. Farm prices continued to decline, and by 1932 the state's manufacturing segment had plummeted. Urban Texans found themselves jobless and rural Texans found themselves getting less for their crops than they cost to grow. In the Panhandle, furthermore, years of drought brought severe dust storms and drifting sand. In Amarillo in 1935, visibility declined to zero a total of seven times as walls of dust shut out the sun. Texas Governor James V. Allred set about bringing federal relief money to Texas, and he succeeded admirably. At its peak, the Civilian Conservation Corps had twenty-seven camps in Texas, working on parks, forests, and other sites. The National Youth Administration employed as many as 18,000 young Texans at a time. The Works Progress Administration and Public Works Administration engaged in a large number of construction projects, including the River Walk in San Antonio. In East Texas, the discovery of a huge oil field in 1930 brought instant riches, as well as chaos, as small towns turned into boom towns. For East Texans who were in the right place at the right time, hard times disappeared in the crush of oil-field development.

The Great Depression failed to daunt Texans' observance of its centennial in 1936. The Texas Centennial celebration employed thousands of Texans between 1934 and 1937, many of them at the state fairgrounds in Dallas. More than seven million people attended the celebrations in Dallas and Fort Worth, which featured the historical pageant "Cavalcade of Texas" and an extravaganza staged by Billy Rose at Casa Mañana. The Texas Centennial brought moments of levity into a situation that continued for many to be grave (Plate 60).

PLATE 57

Geological Map of Texas Showing Approximate Location and Drilling Wells, also Names of Companies Drilling by Standard Blue Print Map and Engineering Company; Fort Worth, 1920. 18" x 18".

Oil was noticed even during the 1543 de Soto Expedition. Commercial exploitation began in 1844 in Navarro County, in the Corsicana oil field. Texas regulated the industry beginning in 1899. The Spindletop discovery was in 1900. The Texas Company, formed in 1902, became Texaco. The J. M. Guffey Petroleum Company, also established in 1902, evolved into the Gulf Oil Company. Humble Oil and Refining, the largest refiner, became Exxon-Mobil. Edna Ferber's novel Giant, *traces wildcatting in West Texas in 1920s. The movie was filmed at Marfa, Texas. The greatest discovery field was in the Permian Basin in West Texas. By 1930 eight major production fields had been discovered. Wildcat boom towns in Winkler, Crane, Howard, and Upton counties followed.*

PLATE 58

State Highway System of Texas by *Dallas Morning News*; Dallas, 1928. 15.5" x 16.5".

The first Texas public thoroughfare was the Old San Antonio Road to Mexico. After the revolution, the new Republic attempted a central national road system, but the effort failed due to poor funding. Better roads were always a concern of both state and local officials, but no coordinated program was adopted until the twentieth century. After the invention of the automobile, "good road" associations lobbied for a statewide transportation network. The State Highway Department came into being in 1917 and by 1921 the Federal government had passed a national matching funds program to encourage states to construct roads. By 1929 Texas had its first system including directional and mileage signage. The system had 30,000 miles of roads, a third of which were hard surfaced. After WWII construction of rural farm-to-market and interstate systems began. Texas contains the largest section of the interstate highway system.

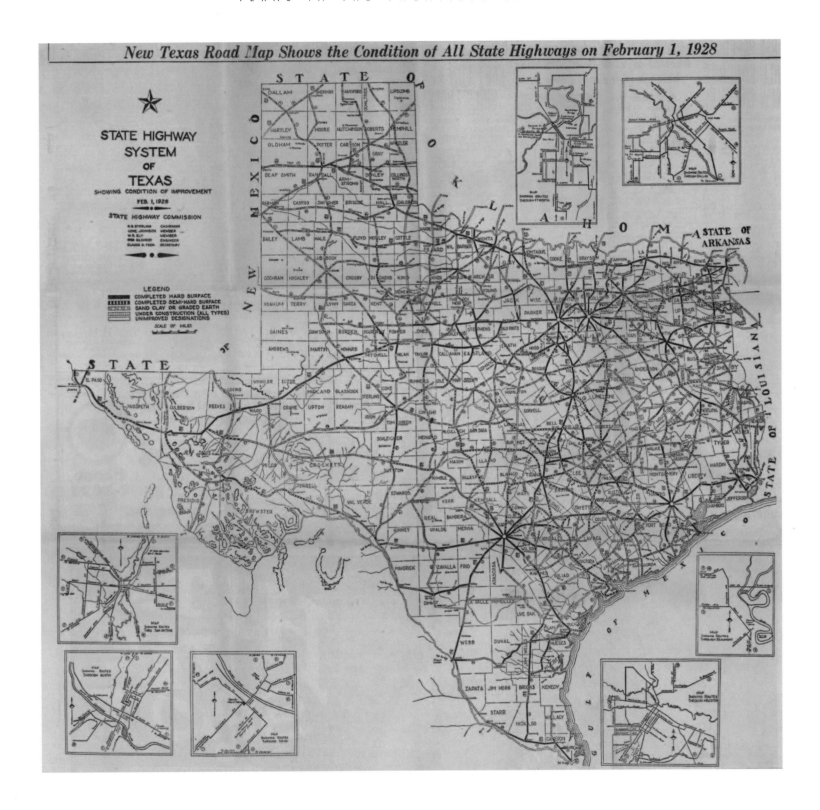

New Texas Road Map Shows the Condition of All State Highways on February 1, 1928

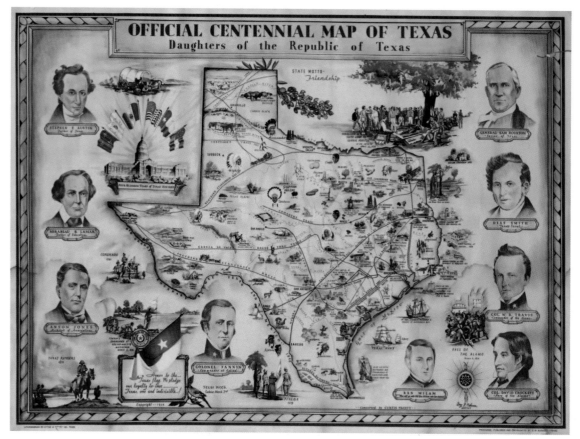

As in the rest of the United States, World War II, not government programs, ended the Great Depression in most of Texas. The changes wrought by the war permanently altered Texas. Military spending imported millions of dollars into the economy, as the state hosted fifteen army bases and forty air bases, beneficiaries of clear, warm weather and abundant land, as well as thirty prisoner-of-war camps. The bases brought both soldiers, some of whom stayed after the war ended, and well-paid jobs for civilians. The state's manufacturing sector increased dramatically, as the area near Houston became the world's largest petrochemical industrial site, and other areas developed steel mills, tin smelters, and pulpwood producers. The state became a major manufacturer of aircraft. To staff these industries, the number of wage earners in the state tripled (Plate 63).

After World War II, most Texans never returned to their rural homes. The industrialization of the state continued apace. During the ten years after 1945, Texas crude oil production doubled, much of it exported through ports on the Gulf Coast. Highway construction throughout the state boomed, and expressways cut through Dallas and Houston. The interstate highway system, originally created to move Cold War troops, began in 1957, and communities rose or fell by their proximity to the great highways. The National Aeronautics and Space Administration placed its new manned-spacecraft center in Houston, near the Ship Channel and the Gulf Freeway, in 1961.

Ordinary Texans took to the skies in increasing numbers during the postwar period. Commercial air service in Texas began before World War II, and Dallas Love Field, Fort Worth Meacham Field, and Houston Hobby Airport all boasted healthy passenger loads. During the 1950s, air passenger traffic rose steadily, with

PLATE 60

Official Centennial Map of Texas, Daughters of the Republic of Texas by C. M. Burnett and Curtis Pruett; Dallas, 1934. Drafting by Guy F. Cahoon. 25" x 35".

This map was commissioned by the Daughters of the Republic of Texas to commemorate the anniversary of state's revolution in celebration of a century of independence. Nine Texas heroes are pictured around the map: Stephen F. Austin (Father of Texas); Mirabeau B. Lamar (Father of Education); Anson Jones (architect of annexation); Colonel James Fannin (commander of Goliad); Ben Milam (Who will go with Old Ben Milam?); Colonel David Crockett (hero of the Alamo); Colonel W. B. Travis (commander of the Alamo); Deaf Smith (Texas scout); General Sam Houston (savior of Texas). Historical events are depicted around and within the outline border of the State. The centerpiece of that celebration was the State Fair Grounds in Dallas. The city won a statewide competition to host the event by pledging the largest amount of cash commitment, twenty-five million dollars in construction funds for fifty buildings. The activity kept Dallas mechanics and draftsmen busy during the Great Depression.

Braniff Airlines leading the way from Dallas. In 1967, Southwest Airlines began regular service between Dallas, Houston, and San Antonio. The largest cities began planning for increased need, with Houston Intercontinental Airport opening in 1969 and the Dallas-Fort Worth International Airport opening in 1973. American Airlines moved its corporate headquarters from New York to Dallas-Fort Worth in 1979. Partially as a result of the growth of the airlines, passenger train service declined, with only two rail lines offering passenger service by 1970. Those lines became part of Amtrak in 1971 and gradually dwindled to only a few passenger trains per week by the mid 1990s.

Continued industrialization brought population growth after World War II and most of the immigrants went to the cities rather than the countryside. By 1960, Texas led the United States in number of urbanized regions, with twenty-one areas of more than 50,000 residents. The suburbs of the core cities began to grow, with populations that were largely young, white, and more affluent than those in the city centers. Those young people reached their jobs in town in cars fueled by petroleum on ever-increasing paved roads. Farmland disappeared under housing developments, a trend that would continue unabated for the next forty years.

The 1960s brought Texas to the forefront of national politics. In 1960, Texas Senator Lyndon Baines Johnson was elected vice president of the United States as the running mate of John F. Kennedy. Three years later, Kennedy's assassination in downtown Dallas brought blame to the state as hurting Americans focused on angry politics that had taken root in Texas. Johnson, winning reelection in 1964, offered an ambitious domestic program aimed at reducing poverty nationwide. At the same time, he greatly increased America's involvement in Vietnam, infuriating many citizens. The furor over the Vietnam War forced Johnson to decide

not to run for reelection in 1968, as he chose instead to retire to his ranch near Stonewall.

While Johnson's Great Society helped bring race to people's attention in the United States, events in Texas had long played a central part in African Americans' fight for desegregation. In 1944, the Supreme Court heard *Smith v. Allwright*, a case filed by the Dallas and Houston chapters of the National Association for the Advancement of Colored People (NAACP), challenging

PLATE 63

American Airlines Route Map by John Fischer. Published by American Airlines in various magazines; Dallas, 1952.

This commercial advertising highlights the ease of airline travel after the war. Point of departure and destination are the primary concerns. The land between is merely "fly over." Ticket examples around the border illustrate all the preparation the seasoned traveler needed to begin a journey. Military aviation, started by General John Pershing during the punitive expedition in 1916, was now an industry. Commercial air traffic began in 1927. The Air Mail Act of 1925 subsidized Texas civilian air carriers. WWII caused over forty military air fields to be constructed statewide. These runways and terminals expanded civilian aviation after the war.

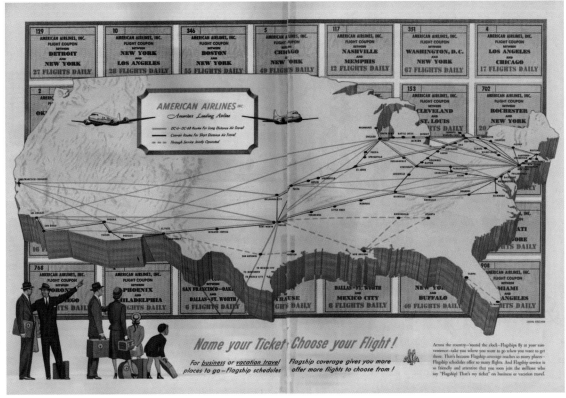

PLATE 59

Cram's Detailed Radio Map – United States and Canada by George F. Cram & Company, Indianapolis. Published by Morton Salt Company, Chicago, 1930. 24" x 33".

United States, Canadian, and Cuban radio stations are shown by call letters and numbers on the Morton Salt Company promotional map. The first Texas radio broadcasts were from the campuses of the universities of Texas at Austin and A&M at College Station in 1911. The City of Dallas

public radio station WRR began in 1920. Two years later, twenty-five Texas commercial stations were in operation including Belo Corporation's WFAA, Dallas and WBAP, Fort Worth. The national broadcast network was statewide by the 1930s, radio's heyday. By mid-century, the first television stations were regularly broadcasting in major Texas cities. These were usually owned by the same companies that controlled the radio stations.

PLATE 64

PLATE 64

Great Military Map of Texas by Joan Kilpatrick and Kenneth Helgren. Published by Texas General Land Office; Austin, 2006.

This is one of the last hand-drafted maps from the Texas General Land Office. Twenty-eight Texas battlegrounds are located, along with seventy-six forts and sixty-two military camps, La Salle's landing and the Glenn Spring raid in 1916. The map pays homage to the courageous men and women who fought for the glory of Texas for almost two centuries.

Texas laws which declared that only white people could vote in primary elections. The Supreme Court found the white primary to be unconstitutional. And in 1950, the court held that the University of Texas could not set up a separate law school for African American Heman Sweatt. *Sweatt v. Painter* desegregated the university's law school and opened the way for African Americans to receive high-quality postgraduate educations. The Civil Rights Act of 1965 ensured the end of *de jure* segregation. Mexican Texans, too, fought for their rights, through organizations such as the Mexican American Legal Defense Fund and the La Raza Unida political party, which fielded a candidate for governor in 1972.

The Texas economy in the last quarter of the twentieth century rose and fell with the boom-and-bust cycles of the oil industry. The dramatic downturn of the 1980s forced a diversification of the economy, with significant growth in the service industries. The computer industry held an important place in Texas, as the silicon transistor was invented at Texas Instruments in Dallas. By 2001, one in four new manufacturing jobs in the state was in computer-related industries, and Dell and Compaq computer companies were based in Texas (Plate 59).

The population of Texas grew steadily throughout the twentieth century, slowing only during the Great Depression. Between 1960 and 2000, the population more than doubled. The federal immigration laws of 1965 created a Texas population more diverse than ever before. With immigration from Asia, Africa, and ever-increasing numbers from Mexico, by 2000, Anglos were no longer the majority of the Texas population. A new Texan might be Indian, Chinese, Thai, Vietnamese, Nigerian, Guatemalan, or from any of the many national groups that migrated to the Lone Star State.

By 1990, Houston, San Antonio, and Dallas ranked among the ten most populous cities in the United States, each ringed with dozens of rapidly expanding suburbs. A number of suburbs, for example Plano, had become large cities in their own right. During the 1990s, Collin County, north of Dallas, doubled in population as the eighth-fastest growing county in the United States. Other counties grew almost as fast, with houses appearing amid cornfields almost overnight.

Texas in the early twenty-first century bears little resemblance to the state one hundred years before (Plate 64). From an agricultural area still dealing with the impact of the Civil War, it became a region of huge cities, filled with people from around the globe. There are still cotton and oil and cattle, but the Texan of today is far more likely to work for a computer company than a cotton broker. We may almost surely presume that the changes of the twenty-first century will be equally breathtaking.

SUGGESTED READINGS

Buenger, Walter L. *The Path to a Modern South: Northeast Texas between Reconstruction and the Great Depression.* Austin: University of Texas Press, 2001.

Johnson, Benjamin H. *Revolution in Texas: How a Forgotten Rebellion and Its Bloody Suppression Turned Mexicans into Americans.* New Haven: Yale University Press, 2003.

McComb, David G. *Houston: A History.* Austin: University of Texas Press, 1981.

Phillips, Michael. *White Metropolis: Race, Ethnicity, and Religion in Dallas, 1841-2001.* Austin: University of Texas Press, 2006.

Turner, Elizabeth Hayes. *Women, Culture, and Community: Religion and Reform in Galveston, 1880-1920.* New York: Oxford University Press, 1997.

INDEX

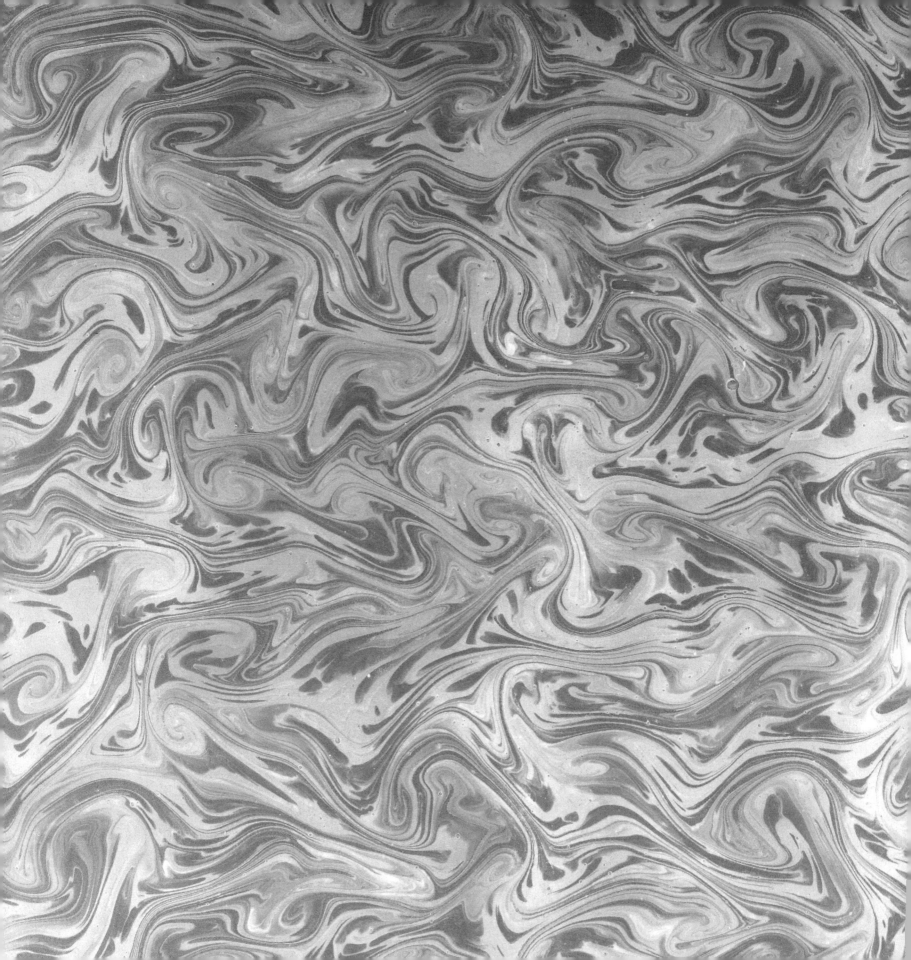